Responsibility in Nanotechnology Development

The International Library of Ethics, Law and Technology

VOLUME 13

For further volumes:
http://www.springer.com/series/7761

Simone Arnaldi • Arianna Ferrari
Paolo Magaudda • Francesca Marin

Editors

Responsibility in Nanotechnology Development

 Springer

Editors
Simone Arnaldi
Centre for Environmental, Ethical, Legal
 and Social Decisions on Emerging
 Technologies (CIGA)
University of Padova
Rovigo, Italy

Istituto Jacques Maritain
Trieste, Italy

Arianna Ferrari
Institute for Technology Assessment
 and Systems Analysis (ITAS)/Karlsruhe
 Institute of Technology (KIT)
Karlsruhe, Germany

Paolo Magaudda
Francesca Marin
Department of Philosophy, Sociology
 Education and Applied Psychology
University of Padova
Padova, Italy

ISSN 1875-0044   ISSN 1875-0036 (electronic)
ISBN 978-94-017-7972-2   ISBN 978-94-017-9103-8 (eBook)
DOI 10.1007/978-94-017-9103-8
Springer Dordrecht Heidelberg New York London

Printed on acid-free paper

Springer is part of Springer Science+Business Media (www.springer.com)

Contents

Part III   Representations and Arrangements of Responsibility

Epilogue: Nanotechnology Beyond Nanotechnologies

Contributors

Simone Arnaldi is Research Fellow of the Centre for Environmental, Ethical, Legal and Social Decisions on Emerging Technologies (CIGA) and of the Department of Political Science, Law, and International Studies at the University of Padova (Italy). He teaches sociology and foresight at the University of Trieste (Italy). He is the Director of the Istituto Jacques Maritain (Italy). His research interests focus on social and media representations of emerging technologies and public engagement in technological decision-making.

Colette Bos is a Ph.D. candidate at the University of Utrecht. She studied Science and Innovation Management at the University of Utrecht and graduated in August 2011. She continued as a Ph.D. in the Innovation Studies group with research within the NanoNextNL program. Her research is conducted under the supervision of Prof. Dr. Ir. Harro van Lente and Dr. Ir. Alexander Peine.

Ilaria Anna Colussi is an Italian attorney and earned a Ph.D. (Doctor Europaeus) in Comparative and European Legal Studies (field: Public Law) from the University of Trento (Italy). Her research interests focus on law, genetics, synthetic biology and new technologies, dealing with the governance of risks and the protection of fundamental rights, from the perspectives of comparative constitutional law and human rights law. She was a member of the Biolaw Project within the Department of Legal Science of the University of Trento, and she has collaborated with the European Centre for Law, Science and New Technologies in Pavia, Italy, since 2009. She was a visiting scholar at the Uehiro Centre for Practical Ethics in Oxford, UK, and at the Interuniversity Chair of Law and the Human Genome in Bilbao, Spain.

Viviana Daloiso graduated in Political Science and earned a Ph.D. in Bioethics, at the "A. Gemelli" School of Medicine (Università Cattolica del Sacro Cuore) in Rome. She is a ranking member of clinical bioethics committees in Rome. Her fields of interest are nanoethics, neuroethics, ethical review of experimental protocols, and forensic bioethics.

Sarah R. Davies is a Marie Curie Research Fellow at the University of Copenhagen. Her Ph.D. (2007) was carried out in Imperial College London's Science Communication

Group; since then, she has worked at Durham University, UK and at Arizona State University's Center for Nanotechnology in Society. Davies has published in journals such as *Science Communication, Science as Culture*, and *Public Understanding of Science*, and has co-edited three volumes (*Science and Its Publics*, 2008; *Understanding Public Debate on Nanotechnologies: Options for Framing Public Policies*, 2010; and *Understanding Nanoscience and Emerging Technologies*, 2010). Her current work explores a number of themes, including materiality and affect in deliberative processes, the governance of emerging technologies, and science communication in museums and science centres.

Arianna Ferrari philosopher, is a researcher at the KIT's Institute for Technology Assessment and Systems Analysis (ITAS) since November 2010. She studied Philosophy at the University of Milan (Italy) and at the University of Tübingen (Germany) and in 2006 completed a Ph.D. in Philosophy (carried out jointly at Tübingen, Germany and Torino, Italy) on ethical and epistemic questions of genetic engineering of animals in biomedical research. Arianna has published widely and lectured on ethical, political and social aspects of emerging technologies, particularly nanotechnologies and genetic engineering. She also teaches and researches on animal philosophy, on the interface between ethics and epistemology as well as ethics and politics in life sciences.

Torsten Fleischer is deputy head of the research area 'Innovation processes and impacts of technology' at the Institute for Technology Assessment and Systems Analysis (ITAS), Karlsruhe Institute of Technology (KIT), and project coordinator for 'Technology Assessment for Nanotechnologies' at ITAS. After graduating in physics, he joined ITAS' predecessor AFAS in 1991. He served as Project Manager for several TA studies for ITAS and the Office of Technology Assessment at the German Parliament (TAB). His working fields are nanotechnology and its implications, new materials and their applications in the energy and transportation sectors, methodological questions of technology assessment as well as the governance of innovation processes and participation therein.

Cecilie Glerup is a Ph.D. student at the Department of Organization, Copenhagen Business School. Her thesis focuses on how political demands for 'responsible innovation' affect both the organization of scientific work in public laboratories and scientists' professional identities. She is furthermore interested in how science and the public interact, and has been involved in the organization of several public engagement events.

Armin Grunwald carried out a Ph.D. in the University of Cologne (Germany) in the field of Theoretical Solid State Physics. After occupations in industry and in the German Aerospace Center, he moved to philosophy and absolved the habilitation at Marburg University. Armin Grunwald holds the Chair of Philosophy and Ethics of Technology at Karlsruhe Institute of Technology (KIT) since 2007. He has been Director of the Institute for Technology Assessment and Systems Analysis (ITAS) at KIT since 1999, and he also has been Director of the Office of Technology

Assessment at the German *Bundestag* since 2002. His main research interests are the ethical aspects of new and emerging sciences and technologies (nanotechnology, synthetic biology, and enhancement technologies), theory and methodology of technology assessment as well as theory and conceptualization of the *Leitbild* of Sustainable Development.

Maja Horst is Head of Department of Media, Cognition and Communication at University of Copenhagen. She holds a masters' degree in communication and a Ph.D. in Science and Technology Studies. Among other projects, she has directed 'Research Management and Risk' (2007–2011) and 'Scientific Social Responsibility' (2009–2014) funded by the Danish Research Council for Social Science. She has published on governance of science and technology, public engagement, science communication and public understanding of science. She has also been experimenting with her own research communication through interactive installations.

Jutta Jahnel food chemist, is a researcher at the KIT's Institute for Technology Assessment and Systems Analysis (ITAS) since July 2010. She studied Food Chemistry and completed her Ph.D. thesis in Chemical Engineering at KIT in Karlsruhe (Germany) on the development of analytical methods in water chemistry. Jutta worked at the DVGW water laboratory and at the Food Control and Animal Health Laboratory in Karlsruhe. At ITAS, she is currently working on environmental, health and safety aspects of nanomaterials with a focus on risk assessment and risk governance frameworks.

Paolo Magaudda is Research Fellow in Sociology at the Department of Philosophy, Sociology, Education and Applied Psychology (FISPPA), University of Padova. In 2011–2012, he was Research Fellow at the Interdepartmental Research Centre for Environmental Law Decisions and Corporate Ethical Certification (CIGA) in the same university. His main research interests are in the fields of science and technology, consumption processes, cultural studies, digital media and popular culture. Among his recent publications are an edited volume (with F. Neresini) on the representations of science and technology on the Italian TV, *La Scienza in TV* (Il Mulino 2011), an ethnographic research on the consumption of musical technologies, *Non solo Oggetti* (Il Mulino 2012) and a book on the representation of science and technology in popular culture, *Innovazione Pop* (Il Mulino 2012).

Francesca Marin philosopher, is Research Fellow at the Department of Philosophy, Sociology, Education and Applied Psychology (FISPPA), University of Padova since February 2013. In April 2011 she completed a Ph.D. in Philosophy at the University of Padova. In 2011 and 2012 she joined the EPOCH project (*Ethics in Public Policy Making: The Case of Human Enhancement*) within the Interdepartmental Research Centre for Environmental Law Decisions and Corporate Ethical Certification (CIGA), University of Padova. Her main research interests include end of life issues, pain management, ethical aspects of emerging technologies and human enhancement. She teaches moral philosophy at the Superior Institute of Religious Sciences (ISSR) in Padova (Italy).

Giuseppe Pellegrini Ph.D. Sociology, teaches methodology of social research at the University of Padova, Italy. His current research focuses on sociology of science, evaluation, citizenship and public participation. He is the coordinator of the research area "Science and Citizens" at the centre Observa Science in Society. His last publications are: *Women and Science: Italy in the International Context* (edited with Barbara Saracino), 2013; *Tecnoscienza, democrazia deliberativa e relazionalità*, Sophia, 1, 105–11, 2013.

Stefanie B. Seitz biologist, has worked for KIT's Institute for Technology Assessment and Systems Analysis (ITAS) since August 2010. She studied Biology at the Friedrich Schiller University of Jena (Germany) and in 2010 completed a Ph.D. in Biology on molecular biologic mechanisms of the circadian clock of the green alga *Chlamydomonas reinhardtii*. Stefanie is currently working on environmental, health and safety aspects of manufactured nanoparticles. Her research interests focus on governance of emerging technologies like nanotechnology or biotechnology including synthetic biology and epigenetics.

Antonio G. Spagnolo graduated in medicine and specialized in Cardiology and in Forensic medicine. He is Full Professor of Bioethics at the "A. Gemelli" Faculty of Medicine (Università Cattolica del Sacro Cuore), Rome, and the Director of the Institute of Bioethics. He is ranking member of several bioethics committees. He is corresponding member of the Pontifical Academy for Life and consultant of the Pontifical Council for Health Care Workers. He is member of the American Nano Society and Member of the Editorial Board of the International Journal of clinical research and bioethics. His fields of interest are: clinical ethics consultation, nanoethics, ethical reviews of experimental protocols, and healthcare ethics committees.

Harro van Lente is Socrates Professor of Philosophy of Sustainable Development at Maastricht University and Associate Professor of Innovation Studies at Utrecht University. He has studied physics and philosophy and has widely published on the dynamics of expectations in science and technology. His research interests concern how emerging technologies – such as nanotechnology, hydrogen and medical technologies – produce novelty and needs. This involves studies of technology assessment, foresight, intermediary organizations, politics of knowledge production and philosophy of technology. Currently, he is Program Director of Technology Assessment of the NanoNextNL, the leading Dutch research consortium in nanotechnology.

Silvia Zullo graduated in Philosophy and earned a Ph.D. in Bioethics at the University of Bologna in 2005, where she is a senior postdoctoral research fellow in bioethics at CIRSFID, Bologna University School of Law. Under several projects supported by the Italian Ministry of Scientific and Technological Research focused on medical genetics, she conducts research on technology, bioethics, and law. Her research takes a legal-philosophical angle, with a specific focus on moral theory and practice in bioethics and law, working on issues relating to the end of life, genetic technologies, and social justice, such as equity, the distribution of risk, and the precautionary principle.

Chapter 1
Introduction: Nanotechnologies and the Quest for Responsibility

Simone Arnaldi, Arianna Ferrari, Paolo Magaudda, and Francesca Marin

1.1 Why Nanotechnology and Responsibility

In the last decade, the field of nanotechnology has changed very quickly from an uncertain promise of benefits and innovations to the ground level of concrete and effective applications. Although still far from science-fiction visions proposed in *Engines of Creation* by Eric Drexler (1986), today nanotechnology has become an actual generator of concrete products and processes, gaining prominence in policy and funding, as well as salience in the public debate and in popular culture over the past few years. Nanotechnology has therefore become an area in which the distance between possibilities and hopes on the one hand, and practical applications on people's lives on the other have been substantial and the connections between these actual presents and possible futures are particularly vague and uncertain, therefore leaving substantial scope to reflect on hypothetical future consequences, even unexpected ones (Selin 2007).

Since the concerns related to nanotechnology's actual and conjectured impacts refer to societal aspects that are very relevant and sensitive, such as the possible consequences on the environment and human health, the quest for a responsible

S. Arnaldi (✉)
Centre for Environmental, Ethical, Legal and Social Decisions on Emerging
Technologies (CIGA), University of Padova, Viale Porta Adige 45, Rovigo, Italy

Istituto Jacques Maritain, Trieste, Italy
e-mail: simone.arnaldi@unipd.it

A. Ferrari
Institute for Technology Assessment and Systems Analysis (ITAS)/Karlsruhe
Institute of Technology (KIT), Karlsruhe, Germany
e-mail: arianna.ferrari@kit.edu

P. Magaudda • F. Marin
Department of Philosophy, Sociology, Education and Applied Psychology,
University of Padova, Padova, Italy

S. Arnaldi et al. (eds.), *Responsibility in Nanotechnology Development*, The International
Library of Ethics, Law and Technology 13, DOI 10.1007/978-94-017-9103-8_1,
© Springer Science+Business Media Dordrecht 2014

development of nanotechnologies has progressively gained momentum in research and policy (cf. Roco 2005; Robinson 2009; McCarthy and Kelty 2010).

A prominent policy example is the launch of the Code of Conduct for Responsible Nanosciences and Nanotechnologies by the European Commission in 2008 (European Commission 2008). The Code aims at enabling safe and beneficial innovation through nanotechnologies and to foster the organization of collective responsibility for the field (von Schomberg 2007). The notion of 'responsible development' works as an overarching ethical framework for innovation, a general foundation of different principles, which should inspire actions (such as sustainability, inclusiveness, excellence, innovation and accountability). The Code was designed to steer responsible research and technology development, so that they should be capable of granting benefits for the society as a whole. In so doing, the Code functions as a instrument for fostering responsibilisation (Dorbeck-Jung and Shelley-Egan 2013), assigning responsibilities to actors and promoting their active involvement, so that cooperation and coordination is strengthened and ensured on a voluntary basis. This assumption of responsibility is described as fundamental for realizing societal goals: the prerequisite here is that the different actors understand and willingly take on the different responsibilities that are connected to the multiple roles in the research and development process they become aware of. In other words, the idea of responsibility in the innovation process in the form of the Code cannot be other than a collective responsibility.

Though prominent, the Commission's Code of Conduct is by no means a unique example and several other instruments attempted to address the junction between responsibility and nanotechnology development, fostering the active commitment of the various actors involved in the field. For example, the voluntary engagement of the relevant social actors was sought through national instruments like the 'Voluntary Reporting Scheme for Engineered Nanoscale Materials', which was promoted by the Department of Environment, Food and Rural Affairs of the UK government in 2006–2008 and was aimed at stimulating an interest by importers and manufacturers of engineered nanomaterials to provide the Department with comprehensive information on material characteristics, as well as with data on toxicity and ecotoxicity (DEFRA 2008a, b). Similarly, the US Environmental Protection Agency (EPA) formally implemented its own voluntary 'stewardship program' for nanoscale materials under the Toxic Substances Control Act (TSCA) in years 2008–2009. Through this voluntary information collection, EPA intended to collaboratively assemble existing data and information from manufacturers, importers, and processors of nanoscale materials in an effort to generate more detailed information of certain specific nanoscale materials. This collaboration between EPA and industry was expected to generate data and analyses for a more complete characterization of materials, and to increase understanding of the environmental health and safety implications of manufactured nanoscale materials for guaranteeing their safe manufacture, processing, distribution, use, storage and disposal (EPA n.d.). At the international level, the OECD Working Party on Nanotechnology (WPN) was established in March 2007 to advise upon emerging policy issues of science, technology and innovation related to the development of nanotechnology and to foster international cooperation that facilitates, among other related issues, the responsible commercialisation of nanotechnology in member countries and certain non-member states (OECD n.d.). In the broader context of the governance of science and technology, the UNESCO Declaration on Bioethics and Human Rights (UNESCO 2005) has affirmed "the desirability of developing new approaches to social responsibility to ensure that progress in science and technology contributes to justice, equity and to the interest of humanity". In general, although it is not nano-specific, the Declaration introduces a "social responsibility principle" (Faunce 2012b), which sets a core group of goals science and technology should be steered to and which is therefore relevant for nanotechnology too. In particular, article 14 of the Declaration lists five "putative public goods" (Faunce 2012b) which the private and public actors involved in science and technology are required to respect: access to quality healthcare and medicines, as well as to nutrition and water; improving of living conditions and environment; elimination of marginalization and exclusion; reduction of poverty and illiteracy. Academic research has proposed the UNESCO Declaration as a "point of departure" for shaping the ethical and human rights principles governing a global project of "artificial photosynthesis", i.e. the replication of photosynthesis, by means of nanoengineering, for localised production of carbon-neutral hydrogen based-fuel and carbohydrate-based food and fertilizer, as forms of planetary therapeutics (Faunce 2012a).

While these examples share the fact that they are initiated by public authorities, private organizations as well launched initiatives for seeking to outline and foster a 'responsible way' to develop nanotechnology. A prominent example is the Responsible Nanocode, which "aims to provide clear guidance about the expected behaviour of companies in relation to their nanotechnology activities" (NIA n.d.) through the implementation of a set of "principles" ranging from "board accountability" (Principle 1) and "worker health and safety" (Principle 3) to "wider social, environmental, health and ethical implications and impacts" (Principle 5). Individual companies like DuPont and BASF developed internal policies, codes of conduct and assessment frameworks for the responsible development of nano-technologies, ensuring safe production, use and disposal of nanoscale materials and identifying, managing and reducing potential health, safety and environmental risks (DuPont 2012; BASF n.d.). Finally, although not "nano-specific", initiatives like ResponsibleCare© for the chemical industry are equally relevant (Heinemann and Schäfer 2009). ResponsibleCare© is aimed to going beyond legislative and regulatory compliance, and by adopting cooperative and voluntary initiatives with government and other stakeholders (ICCA 2006) and commits the "[t]he global chemical industry [to] extend existing local, national and global dialogue processes to enable the industry to address the concerns and expectations of external stakeholders to aid in the continuing development of Responsible Care" (ICCA 2006, 4).

In sum, nanotechnology and responsibility have become a tightly connected pair and policy formulation, academic research, business strategies, and civil society campaigns agree that nanotechnology development should be responsible. Responsibility is not only considered as a value which frames regulation, but as the fundamental condition for enabling good, legitimated and desired technological developments. The transformative power that is attributed to nanotechnology makes this emerging field a perfect candidate to exemplify the consequences of the far-reaching, collective, and uncertain technological endeavour on the notions and practices of responsibility. It is not by chance (cf. Grunwald (Chap. 12) in this book) that this emphasis on the responsible development of nanotechnology has accompanied and sustained over the years the parallel establishment of responsibility as a general feature of technology policy and development. In Europe, such a gradual process has resulted, for instance, in the assumption of the notion of Responsible Research and Innovation (RRI) as a cross-cutting issue under the EU Framework Programme for Research and Innovation "Horizon 2020" (European Commission n.d.; von Schomberg 2013), representing a core value in the new research agenda of the European Union.[1] Similarly, the 'sister concept' of 'responsible innovation' (Owen et al. 2012, 2013) has made its way in the academic debate. Here the idea is that innovation (the new products, services and technologies developed) should not only be simply new, but they should be made and act in the society in a responsible way. These concepts still being in their infancy and despite some differences, they present three shared, distinct features. The first one emphasises the democratic governance of the purposes of research and innovation and their orientation towards the 'right impacts'. The second one values responsiveness, emphasising the integration and institutionalisation of established approaches of anticipation, reflection and deliberation in decision-making processes about research and innovation. The third feature concerns "the framing of responsibility itself in the context of research and innovation as collective activities with uncertain and unpredictable consequences" (Owen et al. 2012). These features are translated in a vision according to which science and society are mutually responsive to each other with a view to the acceptability, sustainability, and societal desirability (von Schomberg 2011).

Responsible innovation is therefore considered an answer to the policy and regulatory dilemmas that are set by techno-scientific fields whose impacts are poorly characterized or highly uncertain. While risk-based governance and the regulatory science that supports it are challenged by the complex and uncertain nature of these phenomena, responsible innovation argues "that stewardship of science and innovation must not only include broad reflection and deliberation on their products, […] but also (and critically) the very *purposes* of science or innovation" (Owen et al. 2013). A discussion on responsibility in nanotechnology development cannot forget this broader context.

1.2 Charting Responsibility: The Structure of the Book

The idea of this book has developed from the acknowledgement that the notion of responsibility is anything but unequivocal and the meanings associated to this notion are extremely diversified in the public discourse of nanoscale technologies.

---

[1] For the parallel development of a distinct notion of 'broader impacts' of research in the US context, cf. Davis and Laas (2013).

Furthermore, these different meanings suggest to commentators and operators different *foci* of attention, ranging from radical appeals to precaution, to the experimentation of new procedures for rule-making, to the implementation of public understanding and/or public engagement activities, and to the development of tests, standards, and measures of exposition for humans and the environment. On the one hand, the formulation and implementation of these policies are affected mostly by our capacity to conjugate what 'responsible development' means for us in the future tense, i.e. with regard to the consequences of our actions on future generations, but also with regard to the assumptions about future situations that influence our way of acting. On the other, assumptions about individuals and their ties to broader social communities affect the solutions for developing nanotechnology responsibly: balancing safety and the legitimate pursuit of knowledge or economic opportunities, individual freedoms and collective interests (in a stronger fashion, the 'common good'), distributing tasks, costs and rewards.

The search for a comprehensive overview of the differentiated concept of responsibility is far beyond the scope of this book, which has the more instrumental goal to chart a landscape of issues, areas and perspectives to examine the current and future configurations of the relationship between nanotechnology and responsibility. Three distinct sections reflecting the multiple levels of the relationship between nanotechnology and responsibility guide the reader in the exploration of these changing notions and practices.

The first section, entitled *Scrutinizing responsibility: theoretical explorations into an entangled concept*, addresses the implications of technological visions for responsibility and examines the criteria and principles that can orient the responsible development of nanotechnology. Focusing on technological visions, Arianna Ferrari and Francesca Marin argue that a different framework is required because the current normative debate on responsibility in new and emerging technologies lacks both explicit acknowledgment of visionary communication about possible technological developments, and awareness of the normative influence of these visions in the present. After offering insights into the etymology of the word 'responsibility' and discussing some examples provided in the literature and regarding the current debate on human enhancement, they show how technological visions shape discourses on emerging technologies and drive research programs as well as our actions and activities. For Ferrari and Marin, thinking about responsibility in relation to technological visions and addressing their normative implications means opening a more fruitful and responsible debate on technological development. Silvia Zullo discusses the contribution of the principle of responsibility and of the precautionary principle to an ethics of responsibility for future generations when faced by policy challenges regarding emerging technologies and their regulation. Zullo stresses the limits of these principles and she argues for the need to integrate utilitarian principles in the equation. From her point of view, the adoption of the principle which demands the maximization of total utility and the principle of maximin in cases where irreversible effects may occur, encourages a concrete intergenerational responsibility, and nurtures a dynamic ethical perspective, both of which are needed to deal with the development of emerging technologies. Moving from a comparison of nanotechnology and synthetic biology, Ilaria Anna Colussi proposes a view of responsibility as a "shared moral obligation" of social actors. Indeed, after presenting similarities and differences between nanotechnology and synthetic biology, Colussi discusses the main principles adopted in the risk analysis model, i.e. the precautionary principle and the proactionary principle, as well as their limits, and finally suggests the notion of 'responsible stewardship'. Being aware of the uncertainty surrounding both risks and benefits of emerging technologies, the proposed model considers alternative actions, immediate and follow-on effects, and interests at stake, letting the technological development go ahead while remaining alert.

The second section on *Responsibility in technology assessment and public engagement* examines the links between the responsible governance of nanotechnology and the practice and mechanisms of technology assessment and public engagement. First, the need to broaden the assessment framework beyond toxicology and beyond scientific experts is discussed. Secondly, the role of Ethics Research Committees in assessing nanotechnology clinical trials protocols is examined. Thirdly, public participation as an instrument for nanotechnology policy is explored. Torsten Fleischer, Jutta Jahnel and Stefanie Seitz point out that the current concept of toxicological risk assessment in the field of nanotechnology (in particular in that referred to manufactured particulate nanomaterials or MPN), which is based on conventional expert-based chemical risk assessment procedures, is too narrow. They start by analysing diverse proposals, such as the one by the International Risk Governance Council based on the considerations of societal impacts and needs, and one for including concerns assessment in the process (concerns of the general public and the stakeholders), which is however still in the early stages. Then, after having discussed the methodological challenges of a broadening of the concept of risk assessment, they discuss the results from a Eurobarometer 2010 as well as from public engagement exercises and focus groups. In the paper the authors call for a wider concept, further developing the idea of concern assessment: this approach should allow for a plurality of actors and different kinds of knowledge which adequately consider societal impacts for understanding risk in a broader sense than in expert-based assessments. Viviana Daloiso and Antonio G. Spagnolo discuss the issue of responsible nanotechnology research in a specific institutional setting: clinical trials and clinical research. According to the Authors, the uncertainty and complexity surrounding the applications of nanotechnologies that are tested in clinical trials assign to Ethics Research Committees (ERCs) for human experimentation the key, if not the decisive, role as public guarantor of the rights and the welfare of trial subjects, while contributing to the increase of available knowledge about human health. In particular, the ERCs must verify that the chosen methodologies are the best suited to the aims of the protocol, that the risk is assessed in terms of probability, magnitude and duration, that the protocol identifies all those elements that may influence that risk. The Authors argue that ERCs' role is even more important in nanomedicine as risks and toxicity change at the nanoscale and that information about them is still not comprehensive. Giuseppe Pellegrini connects responsibility and the public engagement of citizens in decision-making about technologies. Nanotechnology offers a privileged perspective from which to consider the relationship between the development of innovation, ethics and governance, given that the developmental stage of this technology does not allow for a definite characterisation of the main environmental and social issues that are connected to them. The design, production and deployment of nanotechnological innovations can therefore be studied in order to immediately activate pathways of public involvement, even on the basis of similar recent experiences, as in the case of biotechnology.

The third section of the volume on *Representations and configurations of responsibility* deals with some of the ways in which the issue of responsibility in nanotechnology enters the actual processes of innovation and the social discourses about nanotechnology. Colette Bos and Harro van Lente open this section offering a contribution to the current literature regarding corporate social responsibility (CSR) and value chain responsibility (VCR). Given that this literature is particularly focused on existing technologies and value chains, and consequently underestimates firms' views on social responsibility in the light of emerging technologies and new value chains, Bos and van Lente explore these new areas of investigation by presenting three case studies concerning both large and small companies active in the nanotechnology sector. Their empirical results show that changes in the companies' view on social responsibility occur when they deal with new technologies and new value chains. Nevertheless, if the company deals with new technologies but the value chain is stable, then a change in social responsibility is not deemed necessary. In their contribution, Sarah R. Davies, Cecilie Glerup, and Maja Horst point out the contingency and multiplicity of the notion of responsibility by firstly exploring how this concept is articulated within the academic literature. Their discourse analysis conducted on 250 journal articles shows that social responsibility in scientific practice is addressed in two opposing ways: on the one hand, responsibility relies on separating science and society as far as possible; on the other hand, it calls for a greater connection between them. Secondly, a similar diversity arises from the Authors' discussion on how responsibility is performed in the National Science Foundation-funded Center for Nanotechnology in Society at Arizona State University (CNS-ASU) and in the US private sector nano industry with which CNS-ASU sought to interact. While the former performs a broad model of responsible development of nanotechnology, for example by paying attention to its societal dimensions, for the latter responsible development is primarily about ensuring safety. This variety of 'responsibility' both in the literature and in practice calls then for a discussion on what kind of responsibility and responsible development we are looking for. In the following chapter, Paolo Magaudda raises a different perspective about the relationship between responsibility and nanotechnology by focusing of the way responsibility is performed in the actual work of a nanotechnology facility in Italy. In this case, the focus of the analysis is moved to a different perspective, which regards the activation of different forms of mutual responsibility between the actors involved in the work of nanotechnological innovation. Specifically, in the case of innovation performed by a 'boundary organization', we see from the research work of Magaudda that the construction of frameworks of responsibility is linked to at least two aspects: on the one side, to the organizational forms developed to give life to the collective actors emerging during the planning of the research center considered; and on the other, to the strategies and practices of collaboration with other actors, implying the establishment of frameworks of responsibility as well as of distribution of risks and of the construction of regimes of reciprocal trust. In the final chapter of the section, Simone Arnaldi examines the news stories about nanotechnology in the Italian daily press to identify the different representations of responsibility in the coverage. The chapter extends the current research on the definition of responsibility by nanotechnology practitioners and highlights how responsibility is predominantly defined in the terms of the 'traditional contract of science'. This implies that scientists' responsibility is primarily to further scientific knowledge and deliver to society the benefits promised by scientific advances. Also, the analysis shows that the underlying division of labour underlying the 'traditional contract of science' also limits the number and variety of topics on which different social actors can be rightfully considered as sources for the coverage. More specifically, the discussion of radical uncertainties surrounding the nanotechnology enterprises, of precautionary measures, of new institutional arrangements for deliberation on science and technology, is left entirely to civil society organizations, citizens, and humanities scholars.

Finally, an *Epilogue: Nanotechnology beyond nanotechnologies* has the goal to link the discussion on responsibility in nanotechnology development to the broader debate on responsible governance of science, technology and innovation. In this final chapter, Armin Grunwald examines the approach to Responsible Research and Innovation (RRI) and traces back its roots in the debate on nanotechnology. In so doing, the Author shows that the relevance of the debate on nanotechnology and ethics is by no means limited to nanotechnology itself and, instead, it decisively affected the development of a more general 'model' for dealing responsibly with new and emerging sciences and technologies. RRI is presented as an integrative approach to current available instruments to shape science and technology and a multi-fold understanding of responsibility is introduced, which acknowledge three important dimensions (epistemic, empirical and normative). The chapter then examines how, historically, the RRI notion emerged in the context of the nanotechnology debate from the National Nanotechnology Initiative of the U.S. on and how it was then taken up by the European nanotechnology policy. The debate on the Code of Conduct for nanotechnology research and development set in practice by the European Parliament is presented as a landmark in this process and the 'career' of RRI up to the new European research framework programme Horizon 2020 is then recalled. In sum, the chapter shows the parallel development of nano-ethics on the one side, and the debate on Responsible Research and Innovation on the other, thus supporting the view that the emerging debate on the ethics of nanotechnology, as a new and emerging technology promising revolutionary potential but also unclear risk, contributed to the shape of the broader notion of RRI.

1.3   Dealing with an Intractable Object: Perspectives on Responsibility

The overall picture that emerges from this volume reflects the theoretical and empirical diversity of the concept of responsibility. Indeed, by catching and disentangling the different ways in which responsibility can be understood and discussed in nanotechnology development, the concept of responsibility turns out to be complex, multiform and, above all, lacking an univocal definition.

This collection of essays and the tripartite structure described above offer useful entry points to explore the meanings of responsibility, and its junction with nanotechnology. This section of the introduction briefly illustrates three major, horizontal themes that are developed in the essays.

1.3.1 What's in a Name: Responsibility and Social Relationships

Although the different contributions in this book cannot offer a comprehensive picture of all aspects of responsibility, they can be scrutinized to seek (implicit or explicit) similarities and differences in their dealing with definitions and concept building, thus offering useful perspectives for further refinements of this notion.

As a starting point, Grunwald's chapter offers 'a four-place reconstruction [that] generally seems to be suitable for discussing issues of responsibility in scientific and technical progress' (cf. Grunwald). According to this Author, responsibility implies the following elements:

- *someone* (an actor, e.g. a nanotech researcher) assumes responsibility or is made responsible (responsibility is assigned to her/him) for
- *something* such as the results of actions or decisions, e.g. for avoiding adverse health effects of nano-materials, relative to
- *rules and criteria* which orientate responsibility from less responsible or irresponsible action, and relative to the
- *knowledge available* about the impacts and consequences of the action or decision under consideration, including also meta-knowledge about the epistemologic status of that knowledge and the uncertainties involved.

Responsibilities are, therefore, assigned or assumed, thus implying different degrees of active, autonomous commitment of an agent. Assignments and attributions of responsibility affect concrete actors in concrete constellations and are the result of situated social and organizational configurations, which variously connect these four elements. For instance, 'rules and criteria' can define what is relevant as an object of assessment in terms of responsibility ('something'), and what knowledge is relevant for individuals, groups and organizations in such an assessment ('knowledge available'). In turn, the 'knowledge available' can either narrow or broaden what constitutes a consequence (e.g. side effects, long term impacts, etc.), and help define new 'rules and criteria' for responsibility orientation.

Drawing on their discussion of the etymology of the word responsibility (from the Latin word *re-spondeo*, with the two related meanings of 'responding' and 'ensuring'), Ferrari and Marin distinguish four, connected meanings of this notion: (1) responsibility as responding for something, (2) responsibility as responding to someone, (3) responsibility as responding for someone, (4) responsibility as ensuring. In their account, action and its consequences ('something') are central in the discussions about responsibility (*responsibility as responding for something*). Responsibility can be either assigned to or assumed by an agent for her past or future (see below for a development of this aspect), but also, in several cases, for others' action or condition (*responsibility as responding for someone*). Taking responsibility implies the idea of making a commitment to use one's own knowledge, skills, and capacities for ensuring that such a commitment is met (*responsibility as ensuring*). However, no responsibility is possible without a constituency: someone is always responsible (for something, somebody or both) to somebody else, be it a concrete agent (e.g. you, your children, your dog) or an abstraction (e.g. future generations, the people) (*responsibility as responding to someone*). Listening to the needs, desires, questions of others is therefore an undeniable condition of responsibility, because '[a]s a matter of fact, an answer requires both that there is a question and that the content of the question is being listened to' (cf. Ferrari and Marin).

Referring again to Grunwald's four elements, these different forms of responsibility are all assessed against diverse 'rules and criteria' that orient assumption, assignment, and their evaluation. On 'rules and criteria', the chapters in the book adopt different stances, that correspond to two general orientations in the academic debate. On the one hand, several chapters adopt a *descriptive approach* to this aspect, considering requirements and attributions of responsibility to concrete actors in concrete constellations are examined (Bos and van Lente; Fleischer, Jahnel and Seitz; Daloiso and Spagnolo; Pellegrini; Magaudda; Arnaldi; and, partly, Davies, Glerup, and Horst). On the other hand, *normative criteria* for orienting responsible action are sought on a more general level by resorting to utilitarian (cf. Colussi) or other approaches (cf. the 'responsible stewardship model' proposed by Zullo).

The different contributions in this volume can be integrated in a simple, but coherent scheme underlying the understanding of responsibility in the whole book (see Fig. 1.1). Such an understanding places responsibility squarely in the context of social relations, broadly understood, i.e. responsibility has no meaning if it is not a *responsibility to someone* (to be understood as specified above). On a broader level, organizational configurations and policy mechanisms grant institutional force to specific rules and criteria, thus setting boundaries, constraints, and directions for responsible actions (cf. in particular Bos and van Lente; Pellegrini; Magaudda for a reflection on this dimension). Eventually, responsibility is affected by what, in a loose sense, we may call structures, i.e. the material and discursive settings defining science, technology and society relations, which shape the general frame for discussions about responsibility (cf. Davies, Glerup, and Horst; Arnaldi). The contents of responsibility, the dynamics of assumption and assignment, the possibility and conditions of assessment of 'responsible' action are articulated across these three dimensions (see below the next section of the introduction for a development of this topic).

Also, the chapters converge remarkably in treating responsibility in forward-looking terms. The distinction between backward-looking (or retrospective) and forward-looking (or prospective) responsibility has an important place in the definitional

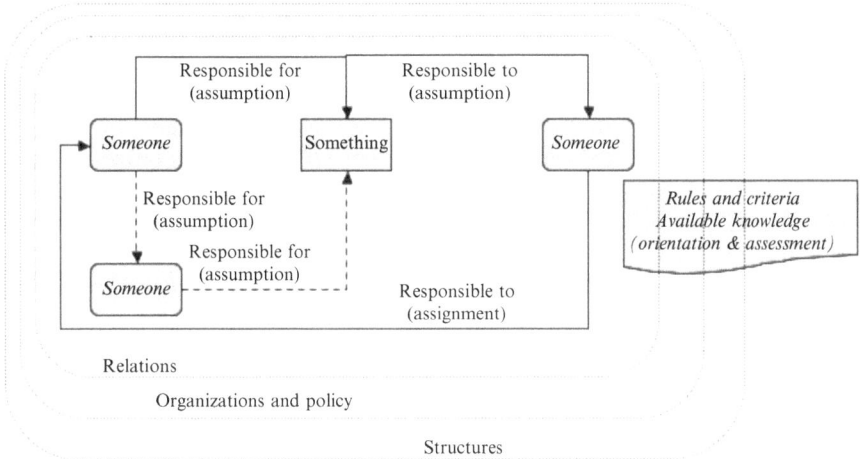

Fig. 1.1 Responsibility and social relations

debates on responsibility in moral philosophy (e.g. Doorn and van de Poel 2012; Vincent 2011) and legal theory (e.g. Gorgoni 2011). To offer a simple definition of both meanings, we draw on and complement Coeckelbergh brief distinction (2012): retrospective responsibility concerns responsibility assignment after something bad (or good) has happened; prospective responsibility regards responsibility assignment in order to prevent bad things happening or to make good things happen. Although what responsibility is for varies across the chapters (e.g. environment, health and security impacts of nanotechnology and synbio for Colussi; collaborative relations and risk management in technology transfer networks for Magaudda), rules, criteria and knowledge are discussed as explicit or implicit conditions to orientate research and technology development, and assessment, in a prospective fashion.

   This emphasis on prospective responsibility over retrospective responsibility is coupled with the emphasis on the assumption of responsibilities for the future over responsibility ascription for past actions and the related the notion of liability as a core meaning of responsibility. Liability (Hart 1968; Young 2006; Vincent 2011) and the related concept of accountability (Davis 2012) imply that to assign moral or legal responsibility 'is to find an agent worthy of a particular kind of reaction; in the case of harm, reactions of blame and perhaps punishment' (Thompson 2012, 205) and, this is the case of 'strict liability', such an adverse treatment can be applied also to cases when the agent did not intend to cause the resultant harm. In line with the emerging views of responsible innovation (cf. Grunwald), this collection of papers emphasises responsibility in nanotechnology as a matter of governing intents and purposes, of science in society, and in science for society (Owen et al. 2012). The next section illustrates how this book illuminates the forms that responsibility assumes in definite situations and contexts.

1.3.2 Situating Responsibility: Division of Labour and Institutional Settings

Theoretical work supports and empirical research confirms the nature of responsibility in social action as the result of an *assignment process* of requirements and attributions of responsibility to concrete actors in concrete constellations (cf. Grunwald in this book). In practical and operative realms, one can always observe a process of 'translation' involving different actors, aims, bonds and opportunities, that are shaped by specific organizational and institutional settings, and by the characteristic of relevant technological applications too.

Fleischer, Jahnel and Seitz connect the topic of responsibility to risk management and risk governance models and emphasize the need to broaden assessment criteria by including societal concerns, which are related to the social perception of risks and technologies, in assessment procedures. Daloiso and Spagnolo show a way through which responsibility in research trials is actually articulated within and by Ethics Research Committees and how it is molded around a specific set of professional and working relationships. Pellegrini points out how in these last few years the distribution of responsibility in technology development has transformed under the pressure to also involve common people and ordinary citizens in the decision-making processes, and not just scientific and political institutions. From different points of view, these contributions demonstrate that the uncertainty surrounding nanotechnology and its impacts demand, and partly have already caused, a change in technology assessment and policy-making configurations, by including new actors (cf. Pellegrini), broadening relevant expertises (cf. Fleischer, Jahnel and Seitz; Daloiso and Spagnolo), changing procedures (cf. Fleischer, Jahnel and Seitz; Pellegrini).

These new or mutating constellations and processes display different forms of coordination. Bos and van Lente consider nanotechnology value chains and observe that coordination and stabilization depend on the stability of the technologies and products that are concerned, as well as on the stability of the number and types of actors that are involved in the value chain. Magaudda shows how boundary organizations affect the definition and distribution of responsibility, according to formal organizational forms and experiential knowledge of other actors' performance. Both chapters introduce trust as a factor influencing the assignment processes that organize collective responsibility.

Such concrete configurations are framed in the more general view of science, technology and society relations and the corresponding conceptions of responsibility. The latter are either reproduced or contested in concrete 'responsible arrangements' and broader scenarios of science-society connections legitimize specific configurations and definitions of responsibility. The findings of Davies et al. for the scientific literature and of Arnaldi for the media coverage demonstrate that conceptions of responsibility depend on the opposing ways in which science-society relations are defined. On the one hand, a discourse of separation and demarcation of science and society supports a view of responsibilities that focus on a narrower focus on the techno-scientific ventures, while society is considered as a beneficiary of technological progress, whose pursuit is considered the primary responsibility of science, together with the creation of science-based technological applications. On the other hand, the acknowledgement of a connection between science and society can be variously framed, from considering society merely as the impacted object of technoscientific advancements to deem the active involvement of citizens and social actors a condition for responsibly shaping science and technology regulation and development.

1.3.3   Responsibility and Orientation to the Future

As this introduction has noted, future-orientation is critical in responsibility. However, this relationship is hardly unproblematic. Considering the collection of essays as a whole, two questions strongly emerge: (1) Which criteria should we follow to responsibly manage the future consequences of our present choices regarding nanotechnology? (2) How does speculation about the future of nanotechnology affect the discourse and practice of responsibility in the present?

Regarding the first question, the book is far from presenting a consensus opinion. Nonetheless, the chapters review some of the major orientations that the literature and policy-making refer to in dealing with this issue, i.e. the precautionary, preventive and proactionary principles. Along with definitions and assessments of these principles, the two chapters by Zullo and Colussi converge in proposing an approach based on 'prudent vigilance', which entails an ongoing evaluation of risks along with benefits, before and after projects are undertaken. Interestingly enough, this common position has different justifications for the two Authors. Zullo's 'tempered utilitarianism' seeks a *via media* between the maximization of the total utility and the maximin principle (see section above for a short definition), which prescribes the ongoing rational assessment of alternative options through the combination of seeking best possible outcomes while avoiding the worst. Colussi's call for a responsible stewardship of nanotechnology is instead based on 'a shared moral obligation' to demonstrate concern for those who are not in a position to represent themselves and for the environment in which future generations will flourish or suffer, and thus act accordingly.

Regarding the second question, future-oriented narratives about nanotechnology (and responsibility) are considered in terms of their performativity. Bos and van Lente examine what happens to responsibility in companies dealing with both new technologies and new value chains. In this case, responsibility is a form of 'responsible speculation', as, within highly speculative value chains, also the considerations about the responsibility within that chain can only remain speculative. On the one hand, as the value chain is stabilized, responsibility assumes specific configurations that differ depending on the extension and heterogeneity of the chain. On the other, speculative views of responsibility correspond to the anticipated images of the company and they can be seen as an attempt to both stabilise the technology and the value chain itself. Ferrari and Marin also consider the performativity of future-oriented narratives, but, instead of focusing on their effect on the definition of responsibility, they examine what responsibility is needed for technological visions. For Ferrari and Marin, taking responsibility for technological visions implies, on the one hand, the critical discussion of the goals and values which frame them and, on the other, the disentanglement of the ways in which these visions act in the present through influencing research programs, scientific agendas and our expectations.

A second, important part of Ferrari and Marin's critique concerns the need for assessing the empirical evidence supporting speculative visions of technology, which are instead often presented as already existing realities and which we have the (moral) responsibility to deal with. This point connects their work with Grunwald's chapter, as this Author identifies the assessment of the status and quality of the knowledge available about the subject of responsibility as one of the three pillars of a discussion on responsible innovation. Along with this epistemological dimension, Grunwald lists an empirical dimension, which refers to behaviour, identities, relations in concrete constellations of actors, and a normative one, which concerns the criteria and rules for judging actions and decisions under consideration as responsible or irresponsible, and for orienting them accordingly.

1.4   Nanotechnology Beyond Nanotechnologies?

The last point to mention is a bit of a paradox. If responsibility is an intractable object, nanotechnology is similarly difficult to circumscribe. There is no point in recalling here the long-lasting discussion about the uncertain boundaries of the field. Throughout all the 'public career' of nanotechnology as a policy and research object, such blurriness has represented a strength to catalyze attention and interest on specific nano-related issues and, at the same time, to set nanotechnology as an exemplary case of innovative S&T governance in the broader context of emerging technologies.

In this volume, the quest for responsibility similarly crosses nanotechnology boundaries. This aspect is particularly evident when other emerging fields or technological applications are considered as reference for elaborating approaches, criteria, and instruments of responsibility (cf. Colussi for synthetic biology and Zullo for neuro-technology and enhancement), as cases sharing similar features (cf. Ferrari and Marin on human enhancement), or as lessons to learn from (cf. Pellegrini on the GMOs controversy).

Conversely, the debate on responsibility in nanotechnology development has the ambition to 'set the tone' for the developing approaches to the responsible governance of emerging technologies as a whole. On this aspect, the chapter by Armin Grunwald outlines a documented and compelling connection between the debate on nano-technology and ethics and the RRI approach. From this point of view, therefore, the considerations and conclusions of this book can offer valuable insights far beyond the perimeter of nanotechnology.

In closing, although the lack of a univocal definition of responsibility might initially be considered as a weakness, it actually allows us to acknowledge the complexity of this concept and to evaluate its different dimensions and aspects. In this way, the logic of innovation and nanotechnology development, which can be understood in terms of actors, impacts and processes, is explored in the three sections and in each of the book's chapters.

Of course, this volume cannot either cover the wide range of meanings and implications related to responsibility or settle the debate on responsibility in nanotechnology development. Nevertheless, our hope is that it can provide a starting point for further discussions and investigations on the way nanotechnology is becoming a pervasive part of our life and what this means for society and our responsibility for the future.

Acknowledgements The Editors wish to thank all the Authors for their competent and inspiring contributions to this volume. Also, we want to thank the three anonymous reviewers for their useful comments on the manuscript. We would like to thank the Centre for environmental, ethical, legal and social decisions on emerging technology (CIGA) of the University of Padova and the Institute of Technology Assessment and Systems Analysis (ITAS)/Karlsruhe Institute of Technology (KIT) for organizing the workshop 'Dilemmas of choice. Responsibility in nano-technology development' (Rovigo, Italy, June 21–22, 2012), from which the initial idea of this book originated. The Editors gratefully acknowledge the financial support of the Fondazione Cassa di Risparmio di Padova e Rovigo for the workshop and of the Istituto Jacques Maritain for the English revision of part of these collected chapters.

References

BASF. n.d. *Nanotechnology code of conduct*. http://www.basf.com/group/corporate/nanotechnol-ogy/en/microsites/nanotechnology/safety/code-of-conduct. Accessed 30 Dec 2013

Coeckelbergh, M. 2012. Moral responsibility, technology, and experiences of the tragic: From Kierkegaard to offshore engineering. *Science and Engineering Ethics* 18(1): 35–48.

Davis, M. 2012. 'Ain't no one here but us social forces': Constructing the professional responsibility of engineers. *Science and Engineering Ethics* 18(1): 13–34.

Davis, M., and K. Laas. 2013. 'Broader impacts' or 'responsible research and innovation'? A comparison of two criteria for funding research in science and engineering. *Science and Engineering Ethics*. doi:10.1007/s11948-013-9480-1.

DEFRA – Department of Environment, Food and Rural Affairs. 2008a. *The voluntary reporting scheme*. Available from: http://archive.defra.gov.uk/environment/quality/nanotech/documents/vrs-nanoscale.pdf. Accessed 30 Dec 2013.

DEFRA – Department of Environment, Food and Rural Affairs. 2008b. *A supplementary guide for the UK voluntary reporting scheme*. Available from: http://archive.defra.gov.uk/environment/quality/nanotech/documents/nano-hazards.pdf. Accessed 30 Dec 2013.

Doorn, N., and I.R. van de Poel. 2012. Editors' overview: Moral responsibility in technology and engineering. *Science and Engineering Ethics* 18(1): 1–11.

Dorbeck-Jung, B., and C. Shelley-Egan. 2013. Meta-regulation and nanotechnologies: The challenge of responsibilisation within the European Commission's code of conduct for responsible nanosciences and nanotechnologies research. *NanoEthics* 7: 55–68.

Drexler, E. 1986. *Engines of creations. The coming era of nanotechnology*. New York: Anchor Books.

DuPont. 2012. *DuPont position statement on nanotechnology*. Available from: http://www.dupont. com/corporate-functions/news-and-events/insights/articles/position-statements/articles/ nanotechnology.html. Accessed 30 Dec 2013.

EPA – Environmental Protection Agency. n.d. *Nanoscale materials stewardship program.* Available from: http://epa.gov/oppt/nano/stewardship.htm. Accessed 30 Dec 2013.

European Commission. 2008. *Commission recommendation of 07/02/2008 on a code of conduct for responsible nanosciences and nanotechnologies research.* Brussels: European Commission.

European Commission. n.d. *Horizon 2020 – The EU framework programme for research and innovation.* http://ec.europa.eu/programmes/horizon2020/. Accessed 30 Dec 2013.

Faunce, T.A. 2012a. Governing planetary nanomedicine: Environmental sustainability and a UNESCO universal declaration on the bioethics and human rights of natural and artificial photosynthesis (global solar fuels and foods). *NanoEthics* 6(1): 15–27.

Faunce, T.A. 2012b. Social responsibility principle. In *Encyclopedia of applied ethics*, vol. 4, 2nd ed, ed. R. Chadwick, 160–166. San Diego: Academic.

Gorgoni, G. 2011. Modelli di responsabilità e regolazione delle nanotecnologie nel diritto comunitario. Dal principio di precauzione ai Codici di Condotta. In *Forme di responsabilità, regolazione e nanotecnologie*, a cura di. G. Guerra, A. Muratorio, E. Pariotti, M. Piccini, and D. Ruggiu, 371–395. Bologna: Il Mulino.

Hart, H.L.A. 1968. *Punishment and responsibility.* Oxford: Clarendon Press.

Heinemann, M., and H. Schäfer. 2009. Guidance for handling and use of nanomaterials at the workplace. *Human and Experimental Toxicology* 28(6–7): 407–411.

ICCA – International Council of Chemical Associations. 2006. *ResponsibleCare® global charter in English.* Available from: http://www.icca-chem.org/ICCADocs/09_RCGC_EN_Feb2006. pdf. Accessed 30 Dec 2013.

McCarthy, E., and C.E. Kelty. 2010. Responsibility and nanotechnology. *Social Studies of Science* 40(3): 405–432.

NIA – Nanotechnology Industries Association. n.d. *Responsible nano-code.* Available from: http://www.nanotechia.org/activities/responsible-nano-code. Accessed 30 Dec 2013.

OECD – Organisation for Economic Cooperation and Development. n.d. *OECD Working Party on Nanotechnology (WPN): Vision statement.* Available from: http://www.oecd.org/sti/nano/ oecdworkingpartyonnanotechnologywpnvisionstatement.htm. Accessed 30 Dec 2013.

Owen, R., P.M. Macnaghten, and J. Stilgoe. 2012. Responsible research and innovation: From science in society to science for society, with society. *Science and Public Policy* 39(6): 751–760.

Owen, R., J. Stilgoe, P.M. Macnaghten, E. Fisher, M. Gorman, and D.H. Guston. 2013. A framework for responsible innovation. In *Responsible innovation*, a cura di. R. Owen, J. Bessant, and M. Heintz, 27–50. London: Wiley.

Robinson, D.K.R. 2009. Co-evolutionary scenarios: An application to prospecting futures of the responsible development of nanotechnology. *Technological Forecasting and Social Change* 76(9): 1222–1239.

Roco, M.C. 2005. Environmentally responsible development of nanotechnology. *Environmental Science and Technology* 39(5): 106–112.

Selin, C. 2007. Expectations and the emergence of nanotechnology. *Science, Technology and Human Values* 32(2): 196–220.

Thompson, A. 2012. The virtue of responsibility for the global climate. In *Ethical adaptation to climate change: Human virtues of the future*, ed. A. Thompson and J. Bendik-Keymer, 203–222. Oxford: Oxford University Press.

UNESCO – United Nations Educational, Scientific, and Cultural Organization. 2005. *Universal declaration on bioethics and human rights.* Available from: http://www.unesco.org/new/en/ social-and-human-sciences/themes/bioethics/bioethics-and-human-rights/. Accessed 31 Jan 2014.

Vincent, N.A. 2011. A structured taxonomy of responsibility concepts. In *Moral responsibility: Beyond free will and determinism*, a cura di. N.A. Vincent, I. van de Poel, and I. van den Hove, 15–25. Dordrecht: Springer.

Von Schombeg, R. 2007. From the ethics of technology to the ethics of knowledge assessment. In *The information society: Innovation, legitimacy, ethics and democracy*, ed. P. Goujon et al., 39–55. Boston: Springer.

Von Schomberg, R. 2011. Prospects for technology assessment in a framework of responsible research and innovation. In *Technikfolgen abschätzen lehren: Bildungspotenziale transdiszip-linärer Methode*, 39–61. Wiesbaden: Springer. https://www.box.com/s/f9quor8jo1bi3ham8lfc. Accessed 30 Dec 2013.

Von Schomberg, R. 2013. A vision of responsible innovation. In *Responsible innovation*, ed. R. Owen, M. Heintz, and J. Bessant, 51–73. London: Wiley.

Young, I.M. 2006. Responsibility and global justice: A social connection model. *Social Philosophy and Policy* 23(1): 102–130.

Part I
Scrutinizing Responsibility: Theoretical Explorations into an Entangled Concept

Chapter 2
Responsibility and Visions in the New and Emerging Technologies

Arianna Ferrari and Francesca Marin

2.1 Introduction

One of the most evident but problematic features of the current normative debate on new and emerging technologies is the lack of explicit acknowledgement of the visionary character of many anticipated applications of technology. Although the ethical debate on existing applications and on applied research in nanotechnology is rather developed (cf. Bos and van Lente in this volume) and focuses mostly on the implications of the risks posed by products containing nanoparticles or other potentially dangerous materials (cf. Fleischer et al. in this volume), the most ethically contentious part of these technologies consists in the applications possible in the distant future as well as in visions. As, for example, MacDonald and Boyce (2008) have pointed out, nanotechnologies have been described as technologies that could potentially provide solutions to problems such as clean, affordable, secure energy (e.g. nanosolar), stronger, lighter, more durable materials (e.g. nanoceramics), low-cost filters to provide clean drinking water (e.g. polymeric nanofiltration), sensors/devices to detect/clean up harmful biological agents or hazardous chemicals in the environment and the means to trigger a revolution in medicine, which will become more 'predictive, preemptive, personalized, and participatory (regenerative)' (cf. Schmidt 2006). Although the ethical literature on nanotechnology acknowledges the nature of this potential and refers to the question of hype (cf., among others, Gordijn 2005), for the most part it continues to discuss these topics in terms of the possible threat to various ethical principles, such as equity, autonomy,

A. Ferrari (✉)
Institute for Technology Assessment and Systems Analysis (ITAS)/Karlsruhe
Institute of Technology (KIT), Karlsruhe, Germany
e-mail: arianna.ferrari@kit.edu

F. Marin
Department of Philosophy, Sociology, Education and Applied Psychology,
University of Padova, Padova, Italy

S. Arnaldi et al. (eds.), *Responsibility in Nanotechnology Development*, The International   21
Library of Ethics, Law and Technology 13, DOI 10.1007/978-94-017-9103-8_2,
© Springer Science+Business Media Dordrecht 2014

privacy, data protection, safety and responsibility (cf., among others, Ebbesen et al. 2006). The technological determinism presupposed in these ethical discussions leads to what has been criticised in the debate as 'speculative ethics' (cf. Nordmann 2007; Nordmann and Rip 2009), a normative reflection detached from a thorough analysis of the state of the art of scientific and technological developments, which leads to biases and problematic discussions (cf. Ferrari et al. 2012).

The same is true of the debate on technologies for human enhancement, a primary topic in this paper. What is meant by the expression 'human enhancement technologies' is, however, far from being clear for two reasons: first, because human enhancement is used as an umbrella term referring to a wide variety of new, emerging and visionary technologies, and second, because the concept of human enhancement is itself controversial and fundamentally normative since it implicitly or explicitly entails the reference to a starting point (which can be a biological status or a parameter), to criteria for judging if something has been raised or not, to the target of the improvement as well as to the subject which realizes it (Grunwald 2008). In this paper we adopt a definition of enhancement originally developed in the STOA[1] project and then used in the EPOCH[2] project. This definition attempts to avoid any intrinsically positive connotation and any dichotomous approach to conceptualizing enhancement interventions and therapeutic interventions. Human enhancement is then taken as referring to any modification aimed at improving individual human performance and brought about by science-based or technology-based interventions in the human body. The use of the technologies in question can be classified on a continuum stretching from non-therapeutic enhancement to restorative therapy (cf. Selgelid 2007).

Talking about the future is something inevitable in the context of the ethics of technology, since scientific programs are always inspired by goals and visions. With Hans Jonas (1979) finally it became clear that the technological and scientific power of human actions extends far beyond the immediate past, having far-reaching implications for future generations and for nature. This is especially the case given the development of biotechnology, by means of which we can change fundamental properties of living beings. With the development of science and technology studies (STS) and technology assessment (TA), it has become increasingly clear that the projections of future scientific and technological developments can be different from what actually occurs. The rise of foresight studies and of the sociology of expectations, which include a range of perspectives such as of the sociology of technology and science, history, economics and innovation studies, has resulted in a description of the influential role that expectations and visions play in shaping the discourse on technology. As Mads Borup and his group have pointed out:

> Such expectations can be seen to be fundamentally 'generative', they guide activities, provide structure and legitimation, attract interest and foster investment. They give definition

---

[1] See Coenen et al. (2009).

[2] Ethics in Public Policy Making: The Case of Human Enhancement (EPOCH) is a European Commission FP7 Science in Society funded project, grant number SIS-CT-2010-266660 (http://epochproject.com).

to roles, clarify duties, offer some shared shape of what to expect and how to prepare for opportunities and risks. Visions drive technical and scientific activity, warranting the production of measurements, calculations, material tests, pilot projects and models. (Borup et al. 2006, pp. 285–286)

Despite various efforts to incorporate the lessons learnt about the complexity and unpredictability of (socio-) technological developments in ethical reflection, we think that the current normative influence of technological visions in the debate has been largely ignored (cf. Ferrari et al. 2012). In this article we want to explore another extension to the consideration of responsibility with respect to the one advocated by Jonas, which comes from disentangling the current role of technological visions in informing experimental research and calling for funding.

We believe that normative analysis referring to the technological future has to be different from just an application of ethical and political theories to a particular technological case, since it should involve considerations on how socioeconomic structures and cultural elements frame the values which inspire technological visions. The normative force of visions in the present is reflected both in how the interaction between technologies and future generations is framed as well as in how the reference to the future is used to justify the allocation of resources in the present. Last but not least, being responsible for the future without falling into the pitfall of technological determinism also means taking responsibility for the values and goals which frame these visions.

After having showed how the current ethical debate has failed to engage with the role of technological visions in the present (Sect. 2.2), we will offer insights into the etymology of the word 'responsibility' necessary for understanding the different normative dimensions of technological visions (Sect. 2.3). Then we will sketch an alternative framework in order to catch the normative issues of new and emerging technologies. We will do this by discussing some examples of the so-called human enhancement technologies (Sect. 2.4). Finally we offer some conclusions (Sect. 2.5).

2.2   A Normative Appraisal of Technological Visions in the Present

What does it mean to be responsible for technological developments which may or may not take place? Although the current approach to conducting ethical studies of the new and emerging technologies includes some important ideas, such as a certain degree of openness toward concrete technological developments and a need to develop a normative analysis as preparatory research, i.e. before it is too late, critical voices have been raised against debates over human enhancement, converging technologies and nanotechnologies, described as leading to impasses and dead-end streets.

Looking at the current debate on nanoethics, for example, we can notice that a large part of it has been reduced to a checklist of the various issues common to other fields of technology, making it in some sense boring, since many issues are simply repeated (cf. Dupuy 2007). Patenaude and his group ask indeed:

> How are we to understand the fact that the philosophical debate over nanotechnologies has been reduced to a clash of seemingly preprogrammed arguments and counterarguments that paralyzes all rational discussion of the ultimate ethical question of social acceptability in matters of nanotechnological development? (Patenaude et al. 2011, p. 285)

A similar dynamic can be seen in the ethical debate over converging technologies. Béland et al. (2011) have talked of an impasse in the ethical debate over Nano-Bio-Info-Cogno (NBIC) for four main reasons. First, any given argument deployed in the debate can serve as the basis for both the positive and the negative evaluation of NBIC; second, it is impossible to provide these arguments with foundations that will enable others to deem them acceptable; third, it is difficult to apply these same arguments to a specific situation; and fourth, the moral argument is ineffective in a democratic society. Although these kinds of discussions are valuable because they reveal different normative positions on the concept of a good life and on the role of technological development in society, they run the risk of staying at a very general level precisely because of the lack of concreteness in describing technologies which do not yet exist.

Much speculation and faith in the technological developments to come are very visible in the current debate on technologies for human enhancement. As Ferrari et al. (2012) have argued in particular for the debate on pharmacological cognitive enhancement (PCE), the lack of thorough study both of the empirical facts around the safety and efficacy of the substances used for cognitive enhancement as well as of the existing data on the social relevance of this phenomenon has led to a problematic ethical debate. Precisely because the empirical evidence for the safety, efficacy and the social relevance of PCE is scarce, an ethical and political discussion of PCE has to be reframed by fully acknowledging the 'visionary' nature of the technological developments being discussed, that is their role as imagined entities and projections in the future that, at the same time, is also active in the present (Ferrari et al. 2012).

The fact that certain technological developments are presented in the literature as being parts of reality at some indefinite point in the future is further complicated by another peculiar character of the new and emerging technologies. Many of them cover very heterogeneous fields, are often characterized by the lack of a common accepted definition (such as in the case of nanotechnologies[3]), or by the fact that sometimes they

---

[3] To the present day there is no commonly shared and general definition of nanotechnology beyond a general identification of the study and control of matter at the molecular and atomic scales (i.e. a definition which gives a precise range or which refers to fields of application). With very few exceptions, it is difficult to find any kind of matter that would not qualify as an object of such nanoscale research: every branch of experimental science and technology nowadays deals with material objects structured at the nanoscale. There are various efforts in different continents to find a definition, which are influenced by the topics regarded as the most important in the local context. In October 2011 Europe adopted the 'Recommendation on the definition of a nanomaterial': 'According to this recommendation, 'nanomaterial' means: 'A natural, incidental or manufactured material containing particles, in an unbound state or as an aggregate or as an agglomerate and

are defined by the goal pursued (such as the case of 'human enhancement') or by the methodological framework used (such as the case of converging technologies). The lack of clarity and heterogeneity in the characterization of new technological fields together with an implicit embrace of technological determinism act to reduce the depth of the analysis of the normative dimensions, which are often reduced to sophisticated risk assessments (cf. Ferrari 2010) or depoliticized in their nature (cf. Felt et al. 2007), and serve as a distraction from concrete questions (such as the allocation of resources).

That similar frustrations and disappointments are present in the ethical debate over different new and emerging technologies is not a coincidence. Indeed, despite their differences from a technical point of view, their goals and visions are deeply interwoven. Nanotechnology is the newest of the converging technologies, at least in the original formulation of NBIC convergence, and is itself multidisciplinary, capable of embodying the perspective of interdisciplinary convergence[4] (cf., among others, Coenen et al. 2004; Saage 2006). In many publications, the ethically contentious part of the idea of converging technologies lies in the goal of human enhancement, as demonstrated by the criticism of the NBIC's framework on converging technologies by their European version (Converging Technologies for the European Knowledge Society – CTEKS; cf. Nordmann 2004). Whereas the credo of NBIC convergence is that we need technological innovation to realize human potential, the credo of CTEKS is, in contrast, that we need social innovation to realize technological potential. Supporters of the NBIC's vision highlight the need for overcoming the bodily and mental imperfections, thus defending the idea of engineering the mind and the body. In contrast, the idea underpinning the European CTEKS's vision is

---

where, for 50 % or more of the particles in the number size distribution, one or more external dimensions is in the size range 1–100 nm. In specific cases and where warranted by concerns for the environment, health, safety or competitiveness the number size distribution threshold of 50 % may be replaced by a threshold between 1 and 50 %. By derogation from the above, fullerenes, graphene flakes and single wall carbon nanotubes with one or more external dimensions below 1 nm should be considered as nanomaterials.' This recommendation states that, by December 2014, the EC will review the definition 'in the light of experience and of scientific and technological developments. The review should particularly focus on whether the number size distribution threshold of 50 % should be increased or decreased.' The absence of a commonly accepted definition of nanotechnologies has precise epistemological implications, because it influences the setting and legitimisation of scientific research areas and therefore the scope of the research. The setting of goals clearly has ethical implications, because goals and aims are shaped by society and because goals are matters of research policy-in particular through priority-setting. The definition of 'nanotechnology' varies depending on research priorities of different countries: (unlike the US, Asian countries such as China, Japan and Korea tend to emphasise material sciences and electronics, while African and Latin American countries focus on environmental sciences and medicine). Furthermore, the lack of a commonly accepted definition is ethically relevant because it opens the ethical discourse indefinitely and, as we will see, many authors tend to associate with nanoethics different kinds of problems (cf. Ferrari 2010).

[4] Since in the project of 'converging technologies' the level of atomic manipulation is taken as the ultimate one and as the basis for creating a new world, the 'integration from the nanoscale' of these technologies is seen as determining a 'tremendous improvement in human abilities and societal outcomes' (Roco and Bainbridge 2002).

that technology adapts the world to the requirements and needs of frail and limited human bodies (engineering for body and for the mind; see Nordmann 2004; Roco and Bainbridge 2002).

Despite the varieties of publications on new and emerging technologies and the fact that utopian and dystopian discourses are acknowledged and analysed historically and culturally (cf. Garreau 2005; Berloznik and Casert 2006; Coenen 2010), the lack of concern for the specific normative dimension of technological visions is striking. The current debate rests on a speculative level and concentrates too much on an indefinite future, because it does not acknowledge either the measures in the present necessary to eventually transform a vision into a reality (such as resources and the criteria of experimental research) or the goals and the values pursued in different research programs. We believe that there is a fundamental difficulty in framing normative analysis, and in particular the analysis of responsibility, in terms of an application of ethical principles to technological visions due to their specific nature. Whereas this method can be fruitful for already existing technologies or technologies in the pipeline, technological visions require a different framework, which starts with the acknowledgement of their specific nature as programs, i.e. as expressions of values and ideas of their promoters which, in order to become concrete, have to pass through a series of steps (from setting up of research programs and different experimental phases including clinical research) (cf. Ferrari et al. 2012).

One main topic around which the debate on human enhancement, and thus on converging technologies and partly on nanotechnologies, has developed through an explicit reference to the future is that of justice. Indeed, it is a matter of projections in the future whether we can argue that some technologies can increase or decrease inequalities among people using them. Whereas the opponents have argued that the introduction of any enhancement technology is that it will create inequality (Annas 2002; McKibben 2003; Fukuyama 2004), proponents have argued that fairness and justice require enhancement, since technologies are there in order to re-establish the inequalities posed by the natural lottery, which randomly distributes capabilities and disabilities (Savulescu 2006, cf. Harris 2007).

But what does it mean in concrete terms to imagine a future society in which some people or even everybody will be enhanced in some way? Although the exercise of imagination is of course not per se something wrong, just as questions regarding the possible impact of the distribution of these technologies are permitted, quite everyone reading this rich literature is left with a sense of dissatisfaction. In their analysis of the discourse, Patenaude and his group have identified seven categories of moral arguments[5] in the debate on human enhancement which appear to be irreducible and lead to an impasse. This irreducibility is due to the presence of different meta-ethical positions, i.e. opinions on the possibility of knowing moral obligations or the human condition as a moral fact and of different conceptions of practical reasoning that correspond to the epistemological positions (Patenaude et al. 2011). These quite general problems in the ethical reflection are exacerbated by two important

---

[5] These arguments are: nature, dignity, the good life, utility, equity, autonomy, and rights (Patenaude et al. 2011).

points which characterize the debate: first, the fact that what is taken as being a human enhancement technology is matter of controversy, and thus something very fuzzy, as already pointed out at the beginning, and second, the fact that in many cases 'human enhancement technologies' do not yet exist, being far-reaching visions and thus not foreseeable in concrete detail.

If we have a look at the examples provided in the literature referring to genetic engineering, we can see that authors engage in major speculation about, for example, enhancing memory by manipulating one's genome. If we have a look at the results of the experiments conducted on animals that are usually quoted in the debate, we can see the materialization of the thesis that empirical facts are not neutral but always subject to interpretation. Julian Savulescu, for example, argues that genetic memory enhancement has been demonstrated in rats and mice and refers to experiments like the 'doogie mice'. But if we accurately analyse the scientific literature, we see that the situation is more complicated: the overexpression of one particular receptor (NMDA receptor 2B) in the precortal frontex of the genetically modified mice has led to an increase in the synaptic activity under electric stimulation, to an improvement of some properties of memory and to a better score on the water maze test. However, these mice have shown different welfare problems, since they suffered from chronic pain, whose origin has not yet been understood, were stressed and have manifested abnormal fear reactions to relatively harmless stimuli, which would constitute a big problem for a life outside the laboratory (Lehrer 2009).

Furthermore, it is still largely unknown how this increase in some properties of memory interrelates to other cognition-related properties. Adding to ethical considerations about the justifiability of such animal experiments as well as to the epistemological problems connected to the transferability of these results to human beings, it is hard to see how these results can be seen as a demonstration of genetic memory enhancement. If we then want to start to make some meaningful analysis of the possible implications of enhancing memory for the distribution of capabilities in a future society, the level of speculation becomes high. Indeed, to do that, we not only need to imagine how the social, economic and political structures of the society would be, but also how a technology would work in very concrete terms.

Assessing whether a modification affects the welfare of a being positively or negatively depends largely on which capabilities are modified and how these capabilities are evaluated. This does not mean, however, that it is not possible to say anything about these experiments. Quite the contrary is true: what has to be discussed is rather which kind of motives and values have led someone to set up the research programs, to pose these questions and to develop arguments pro and contra for the need to ameliorate memory. This becomes something politically very concrete when it comes to initiating a research program. Is a sort of genetic engineering of particular properties of memory on the basis of a poor understanding of cognition-related networks in the brain something we want to pursue at the moment? Who is going to be the subject of such experimentation? Why is it good to allocate money to this research instead of spending money on other purposes? While there is therefore space for considering the normative framework of technological visions, in order to become effective it should be related to questions which link technological visions (the future) with their role in the present. This means exploring the interface between the ethical and political spheres that consider the values which frame current regulatory systems. Together with the ethical principles guiding the allocation of resources and the values guiding the preferences in setting up these visions, these regulatory systems constitute the basis for the introduction of certain technologies.

Appraising this interaction means taking responsibility for the technological visions created. This is a first step toward a fruitful normative analysis of technological visions. However, we can go deeper, firstly by disentangling the different ways in which responsibility can be understood and secondly by showing how this new framework can discover further pathways of reasoning.

2.3   Etymology of the Word *Responsibility*

Investigating and knowing the origins of words can enhance our perspectives about their effective use. For example, by clarifying the original meaning of a term and understanding how it has been transformed over time, we could discover its semantic richness and learn new meanings of the word in question. This kind of investigation is particularly appropriate for a key concept in technological development, namely responsibility, towards which different etymologies have been suggested. Nevertheless, the goal of the following analysis of the word *responsibility* is not to verify the philological correctness of these etymologies, but to explore an alternative account of *responsibility* in the debate on new and emerging technologies.

Although the word *responsibility* is rather modern as it appeared for the first time at the end of the eighteenth century in the context of discussions of *ministerial responsibility*,[6] the adjective responsible has a longer history and it derives from the Latin verb *respondere*, to respond, to give an answer. Even the initial analysis of this verb raises some important considerations concerning the meaning of the term 'responsibility' (Turoldo 2009, p. 7). Indeed, it should firstly be stressed that an answer always follows a certain question or appeal. Quoting Fabrizio Turoldo: 'I respond if someone poses a question, never in general or in the abstract' (Turoldo 2010, p. 174) and 'obviously answering is a consequence of listening' (Turoldo 2010, p. 178). As a matter of fact, an answer requires both that there is a question and that the content of the question is being listened to.

Secondly, answering can occur in different ways because, for example, to 'respond to someone' differs from 'respond for someone'. In addition to listening, the former requires the recognition and acknowledgement of others as well as the

---

[6]According to the Oxford English Dictionary, the first use of the word 'responsibility' is in issue 63 of the 'Federalist' (1787), a text attributed to Alexander Hamilton. For a web version see: http://foundingfathers.info/federalistpapers/fed63.htm, accessed October 2, 2012. As regards the history of the term 'responsibility', see Villey (1977) and Henriot (1977).

respect for individuals and their choices. In this regard it is interesting to note that the term responsibility may also have been derived from the Latin verb *respicere*, which means 'to look back at', 'to gaze at', 'to consider' (Turoldo 2009, p. 8), and then implies acknowledgment of the other person. Furthermore, the word 'respect' has also its origins in *respicere* and consequently, already at the etymological level, a connection between responding to someone and respecting him/her is inevitable.

Unlike the expression 'respond to someone', 'respond for someone' is used in cases involving people who are vulnerable, unable to understand or express their will and need not only listening and respect, but also someone who takes care of them. Moreover, a further distinction can be made between 'respond for someone' and 'respond for something': for example, if a 5-year-old baby (Mary) breaks a fragile object in a store and hurts her hand while her mother (Jenny) is paying the bill, Jenny responds for her daughter and for her injuries (as parents respond for their children). But, if a toy company covertly manufactures toys with dangerous items that endanger children's safety and Jenny buys one of them for Mary, the company responds for injuries to Mary caused by the product because the company responds for the object it has produced.

Consequently, from this first analysis of the word responsibility, we could say that a responsible person is someone who is able to give a response, to answer for something, to or for someone. In this regard, it could be stressed that the link between being responsible and giving an answer is clearly present within the German language as the word *Antwort* (answer) is part of the term *Verantwortung* (responsibility).

Further considerations on this topic could be provided by a deeper analysis of the verb *respondere* because it is derived from *spondere*, another Latin verb meaning 'to promise', 'to guarantee', 'to ensure', 'to make a commitment' (Garcia 1989, p. 51). It may be surprising to note that this verb was used in wedding ceremonies:

> *Spondeo* (I promise) was used in the father's speech in which the father made a commitment to the bridegroom (*sponsus*), giving his daughter (*sponsa*) in marriage in the ceremony of *sponsalia*. The *sponsus* in turn, responded to the father's commitment (*respondeo*), from which responsibility is derived, guaranteeing against the possible uncertainties of the future with a solemn promise (*sponsum*) (Turoldo 2010, p. 180).[7]

As a consequence, following the derivation from *spondere*, *respondere* means not only 'to respond', but also 'to promise something in return' (*re-* 'back' and *spondere* 'to pledge') and one who makes a commitment by means of a promise is then responsible for fulfilling it and should be ready to face the difficulties this may imply.[8]

---

[7] The theme of mutual commitment is also confirmed by the Greek verb *spendo*, which means 'to pour out as a drink offering', 'to make a solemn libation' to the gods, hence 'to engage oneself by a ritual act', but also 'to enter into an agreement' as the gods are called to guarantee an action too (for example the victory in war).

[8] In this respect, it could be noted that the term responsibility may also have been derived from the Latin verb *responsare*, which means 'to be able to go against the mainstream'. Consequently, responsibility is the capacity to face the difficulties that could raise in fulfilling the assumed commitment, going also against the mainstream.

Taking the previous etymological analysis into account and acknowledging the wide range of meanings covered by the word responsibility, it would be reductive to consider responsibility as the ability to assess the results and the consequences of our actions, which is the main meaning generally recognized. Indeed, being described as responding, responsibility calls for listening to the question that needs an answer as well as acknowledgment and respect for others. Moreover, responsibility in the sense of ensuring requires awareness of the assumed commitment with the intent to fulfil it.

2.4   Disentangling Responsibility for Technological Visions

In order to catch the different normative dimensions of technological visions and to highlight their force at the present, in this section we will analyze the main meanings of responsibility addressed above (responsibility as responding for something, responsibility as responding to/for someone and responsibility as ensuring) by discussing some examples provided in the literature and considering the challenges of human enhancement. The methodological choice and the focus on this topic are not accidental. Firstly, we believe that the current ethical debate on human enhancement lacks explicit acknowledgement of the visionary character of many enhancing technological applications. Secondly, we think that this debate underestimates the role of expectations and visions in shaping discourses on these technologies and in driving our actions and activities.

Going back to the development of genetic engineering technologies and in particular to those aimed at enhancing memory through the manipulation of the genome, as already noted above, advocates use scientific results in a very easygoing way just to support their political research agenda. However, this is problematic because it skips the reference to technologies (in this case the reference to some scientific experiments) as expressions of research visions and not as realities. Since the debate on this topic generally refers to the results of experiments conducted on animals and/or to human enhancement technologies that do not yet exist, there is a responsibility in the sense of to respond for something that is a responsibility for the created technological visions and for their impact on scientific research. Indeed, the creation of these visions leads to certain research programs being set up, which calls for a justification of the allocation of these resources in these specific areas of research.

Furthermore, as pointed out above, an answer always follows a certain question or appeal and is a consequence of listening. Consequently, being responsible for technological visions means that social requirements have been taken into consideration, and that such visions translate society's interests. In this sense, the meaning of responsibility as responding to someone is at stake because, for instance, technological visions inasmuch as they provide orientation for research, need to be based either on robust empirical data or on broad social desires or on both, if they do not want to be just the expressions of particular interests which use speculation to push research agendas. The lack of reference to these kinds of data has been pointed out as a problem in particular in the debate on pharmacological cognitive enhancement (cf. Outram 2011; Quednow 2011; Racine and Forlini 2010). As argued by Ferrari et al. (2012), the tendency towards using stimulant drugs for the purpose of cognitive enhancement appears to be greater among academics and students, namely among social groups that consider cognitive capacities particularly relevant in their work. In this way, the technological visions related to cognitive enhancement respond to the desires of members of a social minority group, which calls for the motives which drive technological development in this direction rather than in another to be justified.

Moreover, responding for something presupposes responding for someone. Going back to the example provided in the previous section, the toy company that manufactures toys with dangerous items responds not only for the object it has produced, but also for potential injuries to people caused by the product. Similarly, given that the technological visions act in the present and have an impact on the social context by driving our actions and activities, responding for these visions also means responding for the social, economic and political structures of society. For example, inasmuch as technological visions encourage an improvement in our abilities and capabilities, modifications of our social relationships toward disabled people or vulnerable persons take place that may, for instance, marginalize them or preserve architectural barriers. In this sense, human capabilities are evaluated in quantitative terms rather than in qualitative ones, and a discussion of these issues and of the reasons for the requested human enhancement appears to be scarce, or even absent, in the ethical debate on this topic.

Considering another example of human enhancement technologies, neuroenhancement in individuals under 18 years of age is often considered in the literature as a growing phenomenon. For Ilina Singh and Kelly J. Kelleher this phenomenon is socially relevant and will become more and more common with the availability of psychotropic drugs and pharmacological neuroenhancers (including methylphenidate, e.g. Ritalin, and dexamphetamine compounds, e.g. Adderall):

> It is clear from a number of reports that neuroenhancement is actively practiced by adults and by young people in North America, Europe, and the United Kingdom [...] Informal polls and newspaper reports suggest that alongside growing evidence of neuroenhancement practices, there is also increased public tolerance of neuroenhancement using psychotropic drugs-at least among educated middle-class respondents [...] We think the current level of use of stimulants for purposes of enhancement among young people (which is almost certainly underreported) will also increase, not least because the use of psychotropic neuroenhancing agents will likely become normal in future generations. (Singh and Kelleher 2010, p. 3)

Furthermore, from Singh and Kelleher's point of view, although at present these psychotropic drugs are almost always prescribed to children and/or adolescents diagnosed for attention deficit-hyperactivity disorder (ADHD), there are suspicions surrounding the parallel increases in ADHD diagnoses and in stimulant drug prescriptions.

> Stimulants have been shown to improve focus and attention in young people without a psychiatric diagnosis as well as in young people with ADHD [...] Consequently, there is widespread suspicion that the simultaneous increases in ADHD diagnoses and in stimulant drug prescriptions reflect in part an increasing use of stimulants to enhance young people's performance, rather than to treat a clearly diagnosed disorder. (Singh and Kelleher 2010, p. 5)

These considerations exemplify the fact that technological visions are not discussed *qua* visions but that they are taken for granted. Indeed, in Sect. 2.2 of this paper we noted that the empirical evidence for the safety, efficacy and social relevance of cognitive enhancement is scarce. Moreover, this kind of enhancement is presented in literature as an existing reality which must be regulated. As a matter of fact, shifting from the considerations addressed above, in their contribution Singh and Kelleher suggest specific research, practice and policy recommendations concerning the use of stimulant drugs for neuroenhancement in young people (Singh and Kelleher 2010, pp. 7–13).

What is missing in this discussion is the fact that the assumption that the growing use of psychotropic drugs and pharmacological neuroenhancers among young people is likely to be inevitable and that neuroenhancement is an increasingly common practice means implicating both practitioners in the medical field and parents. As a matter of fact the prescription of drugs for the purpose of neuroenhancement occurs within the context of a health professional-children-parents relationship. The relationship is always described in terms of an inequality of knowledge and skills because it is established between a person seeking assistance and another who professes to provide it. This understanding is suggested by the etymology of the word 'profession,' because it is derived from the Latin verb profiteri, which means 'to declare aloud or publicly.' The health professional declares that by using his knowledge and skills he will provide the necessary help to promote the interest of someone seeking assistance. As a consequence, the health professional is not a simple technician, but one who is involved in a particular human relationship making a professional commitment towards others that do not possess his knowledge and skills. Responsibility in the sense of ensuring is thus at stake, and it is extremely relevant concerning the prescription of stimulant drugs for neuroenhancement in young people, primarily for two reasons: (1) the uncertain decision-making capacity of children, adolescents and teens, and (2) the scarce empirical evidence for the safety, efficacy and the social relevance of cognitive enhancers. Within the health professional-children-parents relationship, responsibility in the sense of responding for someone is at stake as well because parents should respond for their child and for his/her possible request for neuroenhancement, thoroughly assessing the reasons behind this request and eventually questioning their conduct.

Another aspect of the meaning of responsibility in the sense of responding for someone concerns responding for future generations. In this respect, a paradigmatic example showing the normative force of technological visions in the present can be found in the debate on genetic enhancement, in particular in the normative theory proposed on this topic by Savulescu. Adopting an utilitarian approach, he argues that what matters is the promotion of human well-being, described as 'the very essence of what is necessary for a good human life' (Savulescu 2007, p. 530), and claims that enhancement promotes well-being because it increases the chances of leading a good/better life (Savulescu 2009, p. 222; cf. Savulescu et al. 2011). For Savulescu, our existence is better/good when it promotes the maximization of capabilities and the minimization of disabilities. Furthermore, in his opinion, enhancing our children by employing genetic engineering technologies is not only permissible but also a matter of 'moral obligation or moral reason',[9] such as in treating and preventing diseases. Given that genetic enhancement interventions can improve one's native equipment and increase the opportunities for one to lead a better life, in Savulescu's opinion we should promote well-being by means of enhancement. By affirming that we have a moral reason to enhance our children, he means that in the absence of some other reason for action, a person who has a good reason to improve his offspring is morally required to have the best child possible.

We believe that the normative approach proposed by Savulescu attests the visionary character of genetic enhancing technologies and its relevance not only in driving actions, but also in clarifying and/or creating moral duties. Indeed, although these technologies do not yet exist, at stake in the debate over affirming that genetic enhancement promotes human well-being are expectations and visions that have inevitable ripple effects into science's agenda, public perception and resource allocation. The parents' role is also implicated in this normative framework, by arguing for the existence of a moral duty to have the best child possible, which then become a duty to use these enhancing technologies. Furthermore, within Savulescu's approach, responding for future generations raises a problematic issue, because to genetically enhance offspring means imposing the parents' idea of well-being on their child, inscribing it in his or her own biological nature. Savulescu considers this when he asserts that reproductive choices must be based on 'a plausible conception of well-being and better life for the child' (Savulescu 2007, p. 527). Nevertheless, the concept of well-being is ambiguous, and the alteration of the gene pool cannot be carried out in accordance with the offspring's concept of well-being.

2.5 Conclusions

The aim of this paper was to offer an alternative framework for addressing the normative implications of technological visions, in particular those referring to human enhancement. We believe that the current ethical debate on new and emerging technologies is trapped in an impasse because it lacks any acknowledgement of the visionary character of the technologies at stake and of the current normative influence of these technological visions. Indeed, as shown by the examples discussed above, these visions drive our actions, modify our roles (as parents, as citizens, as professionals, as policy-makers etc.), and influence research programs, scientific agendas and our understanding of our moral duties. By shaping our expectations, technological visions also influence our perception of the ethical issues at stake, for example by concentrating the debate on the legitimacy of modifying memory abilities in general instead of addressing interrelations with other cognitive abilities or the need to test devices and drugs on healthy subjects. Therefore we believe that a

---

[9] It should be noted that the expressions 'moral obligation' and 'moral reason' are not synonymous because the former expresses a stronger normative force than the latter. For a critical analysis of the use of 'have a good reason to' within Savulescu's approach, see Marin (2012, pp. 113–115).

normative analysis should move from the question of meaning to disentangling the reasons and the values behind these visions.

This approach also encompasses the wide range of meanings covered by the word 'responsibility': responsibility as responding for something, as responding to/ for someone and as ensuring. Nevertheless we are aware that it remains difficult to identify the subjects who are responsible for particular technological visions and for their impact on society. This points to a limit of ethical reflection and at the same time to the need for exploring the interface with the political dimension, since the reasons for the desirability of technological visions should be investigated in a particular sociopolitical (and cultural) context.

However, we also think that the suggested alternative framework opens a more fruitful course of debate on technological development, because paying attention to the different steps required to accept or refuse and to render a technological vision a reality makes it possible for us to grasp the main normative issues at stake and thus to conduct the debate in a responsible way.

References

Annas, G. 2002. *Cell division*. Center for Genetics and Society. http://www.geneticsandsociety. org/article.php?id=164. Accessed 21 Oct 2012.

Béland, J.P., J. Patenaude, G.A. Legault, P. Boissy, and M. Parent. 2011. The social and ethical acceptability of NBICs for purposes of human enhancement: Why does the debate remain mired in impasse? *NanoEthics* 5(3): 295–307.

Berloznik, R., and Casert, R. 2006. *Technology assessment on converging technology – Literature study and vision assessment*. Background document for the STOA workshop 'converging technologies in the 21st century: Heaven, hell or down to earth?', Brussels, 27 June 2006.

Borup, M., N. Brown, K. Konrad, and H. van Lente. 2006. The sociology of expectations in science and technology. *Technology Analysis and Strategic Management* 18(3/4): 285–298.

Coenen, C. 2010. Deliberating visions: The case of human enhancement in the discourse on nanotechnology and convergence. In *Governing future technologies. Nanotechnology and the rise of an assessment regime*, ed. M. Kaiser et al., 73–88. Dordrecht: Springer.

Coenen, C., T. Fleischer, and M. Rader. 2004. Of visions, dreams, and nightmares: The debate on converging technologies. *Technikfolgenabschätzung – Theorie und Praxis* 13(3): 118–125.

Coenen, C., M. Smits, M. Schuijff, P. Klaassen, L. Hennen, M. Rader, and G. Wolbring. 2009. *Human enhancement study* (IP/A/STOA/FWC/2005_28/SC32 and 39). Brussels: European Parliament. http://www.europarl.europa.eu/RegData/etudes/etudes/join/2009/429976/IPOL-JOIN_ ET(2009)429976_EN.pdf. Accessed 24 Jun 2014.

Dupuy, J.-P. 2007. Some pitfalls in the philosophical foundations of nanoethics. *The Journal of Medicine and Philosophy* 32(3): 237–261.

Ebbesen, M., S. Andersen, and F. Besenbacher. 2006. Ethics in nanotechnology: Starting from scratch? *Bulletin of Science Technology Society* 26(6): 451–462.

Felt, U., et al. 2007. *Taking European knowledge society seriously*. Report of the expert group on science and governance to the science, economy and society directorate, Directorate-General for Research, European Commission. http://ec.europa.eu/research/science-society/document_ library/pdf_06/european-knowledge-society_en.pdf. Accessed 21 Oct 2012.

Ferrari, Arianna. 2010. Developments in the debate on nanoethics: Traditional approaches and the need for new kinds of analysis. *NanoEthics* 4(1): 27–52.

Ferrari, A., C. Coenen, and A. Grunwald. 2012. Visions and ethics in current discourse on human enhancement. *NanoEthics* 6(3): 215–229. doi:10.1007/s11569-012-0155-1.

Fukuyama, F. 2004. Transhumanism. *Foreign Policy* 144: 42–44.

Garcia, D. 1989. *Fundamentos de bioetica*. Madrid: EUDEMA.

Garreau, R. 2005. *Radical evolution. The promises and perils of enhancing our minds, our bodies – And what it means to be human*. Toronto: Doubleday.

Gordijn, B. 2005. Nanoethics: From utopian dreams and apocalyptic nightmares towards a more balanced view. *Science and Engineering Ethics* 11(4): 521–533.

Grunwald, A. 2008. *Auf dem Weg in eine nanotechnologische Zukunft: Philosophisch-ethische Fragen*. Freiburg: Karl Alber.

Harris, J. 2007. *Enhancing evolution*. Oxford: Oxford University Press.

Henriot, J. 1977. Note sur la date et le sens de l'apparition du mot responsabilité. *Archives de Philosophie du Droit* 22: 58–62.

Jonas, H. 1979. *Das Prinzip Verantwortung: Versuch einer Ethik für die technologische Zivilisation*. Suhrkamp: Frankfurt am Main.

Lehrer, J. 2009. Neuroscience: Small, furry…and smart. *Nature* 461(7266): 862–864.

MacDonald, L., and G.J.S. Boyce. 2008. Nanotechnology: Considering the complex ethical, legal, and societal issues with the parameters of human performance. *NanoEthics* 2(3): 265–275.

Marin, F. 2012. *Il bene del paziente e le sue metamorfosi nell'etica biomedica*. Milano: Bruno Mondadori.

McKibben, B. 2003. *Designer genes*. Orion, April 30. http://www.orionsociety.org/pages/om/03-3om/McKibben.html. Accessed 21 Oct 2012.

Nordmann, A. 2004. *Converging technologies – Shaping the future of European societies*. Report of the HLEG Foresighting the New Technology Wave. http://www.ntnu.no/2020/final_report_en.pdf. Accessed 21 Oct 2012.

Nordmann, A. 2007. If and then: A critique of speculative nanoethics. *NanoEthics* 1: 31–46.

Nordmann, A., and A. Rip. 2009. Mind the gap revisited. *Nature Nanotechnology* 4(5): 273–274.

Outram, S.M. 2011. Ethical considerations in the framing of the cognitive enhancement debate. *Neuroethics* 5(2): 173–184.

Patenaude, J., G.A. Legault, J.P. Béland, P. Boissy, and M. Parent. 2011. Moral arguments in the debate over nanotechnologies: Are we talking past each other? *NanoEthics* 5(3): 285–293.

Quednow, B. 2011. Ethics of neuroenhancement: A phantom debate. *BioSocieties* 5(1): 153–156.

Racine, E., and C. Forlini. 2010. Cognitive enhancement, lifestyle choice or misuse of prescription drugs? Ethics blind spots in current debates. *Neuroethics* 3(1): 1–4.

Roco, M., and W.S. Bainbridge (eds.). 2002. *Converging technologies for improving human performance: Nanotechnology, biotechnology, information technology and cognitive science*. Arlington: National Science Foundation.

Saage, R. 2006. Konvergenztechnologische Zukunftsvisionen und der klassische Utopiediskurs. In *Nanotechnologien im Kontext*, ed. A. Nordmann, J. Schummer, and A. Schwarz, 179–194. Berlin: Akademische Verlagsgesellschaft.

Savulescu, J. 2006. Justice, fairness, and enhancement. *Annals of the New York Academy of Sciences* 1093: 321–338.

Savulescu, J. 2007. Genetic interventions and the ethics of enhancement of human beings. In *The Oxford handbook of bioethics*, ed. B. Steinbock, 516–535. Oxford: Oxford University Press.

Savulescu, J. 2009. Genetic enhancement. In *A companion to bioethics*, ed. H. Kuhse and P. Singer, 216–234. Oxford: Blackwell.

Savulescu, J., A. Sandberg, and G. Kahane. 2011. Well-being and enhancement. In *Enhancing human capacities*, ed. J. Savulescu, R. ter Meulen, and G. Kahane, 3–18. Oxford: Wiley-Blackwell.

Schmidt, K.F. 2006. *Nanofrontiers: Visions for the future of nanotechnology*. Washington, DC: Project on Emerging Nanotechnologies.

Selgelid, M. 2007. An argument against arguments for enhancement. *Studies in Ethics, Law, and Technology* 1(1): 1–7. doi:10.2202/1941–6008.1008.

Singh, I., and K.J. Kelleher. 2010. Neuroenhancement in young people: Proposal for research, policy, and clinical management. *AJOB Neuroscience* 1(1): 3–16.

Turoldo, F. 2009. *Bioetica ed etica della responsabilità: dai fondamenti teorici alle applicazioni pratiche*. Assisi: Cittadella Editore.

Turoldo, F. 2010. Ethics of responsibility in a multicultural context. *Perspectives in Biology and Medicine* 53(2): 174–185.

Villey, M. 1977. Esquisse historique sur le mot responsible. *Archives de Philosophie du Droit* 22: 45–58.

Chapter 3
Features of Intergenerational Moral Responsibility in the Age of the Emerging Technologies

Silvia Zullo

3.1 Nanotechnologies and Human Enhancement: Ethical and Social Challenges

Scholars and experts predict that innovation in nanotechnology and all emerging technologies will be raising serious ethical and social issues even in this decade.[1] On the one hand, these same technologies could create novel opportunities by offering new kinds of therapies for treating what are now mortal diseases, such as cancer and Parkinson's disease. And the advancements made in nanomedicine could help many people in maintaining their health and independence and in having an active role in society. But on the other hand, these very advancements bring with them a host of problems, a notable example of which—combining nano and gene applications—lies in human enhancement, that is, the use of nanotechnologies on human cells to prevent them from aging, or to augment their capacities, or to restore lost abilities. This last use is clearly less contentious. But we can see a future in which the use of genetic interventions will give rise to issues of social justice and equal opportunity, thus bringing into play the role of social institutions: can we meaningfully distinguish between treatment and genetic enhancement and, if so, should enhancement be subsidized in the same way that treatment is, or should it be left to individual choice and means?

---

[1] Nanotechnology is defined by the National Nanotechnology Initiative as 'the understanding and control of matter at dimensions of roughly 1–100 nm, where unique phenomena enable novel applications. Encompassing nanoscale science, engineering and technology, nanotechnology involves imaging, measuring, modelling, and manipulating matter at this length scale. At the nanoscale, the physical, chemical, and biological properties of materials differ in fundamental and valuable ways from the properties of individual atoms and molecules or bulk matter. Nanotechnology is directed toward understanding and creating improved materials, devices, and systems that could use these new properties.'

S. Zullo (✉)
Interdepartmental Center for Research in the History, Philosophy, and Sociology
of Law and in Legal Informatics (CIRSFID), University of Bologna, Bologna, Italy
e-mail: silvia.zullo@unibo.it

S. Arnaldi et al. (eds.), *Responsibility in Nanotechnology Development*, The International Library of Ethics, Law and Technology 13, DOI 10.1007/978-94-017-9103-8_3,
© Springer Science+Business Media Dordrecht 2014

One criterion on which basis to treat this problem is the one put forward by Ruth Chadwick, the idea that 'morally permissible enhancements are those which reduce or at least do not increase social inequalities' (Chadwick 2008). So, considering that society's resources are limited, and not all of its resources can be allocated to the *naturally* disadvantaged, since we already have to take care of the *socially* disadvantaged, we need to decide how many resources (or what costs) society is willing to devote to human enhancement of the naturally disadvantaged and what amount of benefits is acceptable. This problem is amenable to quantitative interpretation: to enhance x means to add to, exaggerate, or increase x in some respect, so we need to be specific about the respect in which x is enhanced. This way of looking at the issue means that interventions need to be assessed on a case-by-case basis rather than be judged in general terms.

If gene therapy can be used to restore normal functioning for handicapped individuals, the cost of those procedures must be taken in account. Furthermore, individuals with greater resources can pay for their own enhancement, but this creates social inequalities for those who cannot. We can bring into play here Rawls's (1971) principle of equality of opportunity, but this principle requires only that opportunities be equal for people who are similarly talented and motivated, so the normal range of opportunity will not be equally distributed. It seems that the argument for an adequate minimum threshold of opportunities cannot be justified if it requires an absolute equality of genomes for all. The problem is unsolved because we need to define what is treatment and what is not. Furthermore, the idea of treatment as treatment designed to ensure normal functioning in a post-genomic society is very difficult, because medicine has changed its role (from restoring to 'improving' health), and also because there are sociocultural pressures that condition the individual to perceive normality in a subjective and distorted way.[2]

It is possible that future genetic enhancement will make individuals radically different from one another: if so, this change could negatively impact the reciprocity in moral relationships, making unenhanced people less competitive in society and leading to distinctions between genetic under- and upper classes.[3]

The same concerns emerge in the discussion on nanotechnologies and nanomedicine: while many applications are indeed desirable (an example might be treatment for a brain tumor), others may be undesirable and fraught with risks for proper cognitive function. In fact, technological advancements in neuroscience involve risks arising from so-called smart drugs (nonmedical uses of psychopharmaceuticals) or from inappropriate uses of functional neuroimages (see Savulescu and Persson 2008).

---

[2] This situation not only favours an idea of normality perceived on the basis of subjective reasoning (for example, feeling disadvantaged in terms of intelligence), but also reframes the idea of the objective purposes of medicine, an idea that in a postgenomic society turns into 'treatment on demand.' So, in the future, those who fall short of some technically achievable ideals would increasingly be seen as 'marginalized,' and the standards for what is genetically desirable will be dictated by economically and politically dominant groups. These are implications of genetic enhancement that a politically and economically liberal society must now face, since genetic enhancement goes beyond the purposes of relieving pain and the cure of disease.

[3] On the way enhancement, morality, fairness, and theories of justice relate to the concept of health, see Orlebeke Caldera (2008), Lindsay (2005), Schermer (2011), and Levy (2011).

Clearly, the medical use of nanoparticles will open the door to new futuristic scenarios. For example, scientists agree that not too long from now we will be able to use nanoimages to study the beginning and progress of a disease, and we will also be able to build new functional molecules to empower the human immune system.

If on the one hand these new therapeutic possibilities could help us control neurogenerative, cardiovascular diseases and cancer, on the other hand they also invites a certain prudence as concerns their use in the biomedical field. Indeed, we should not underestimate the risks that may arise once the human body interacts with nanoparticles. These risks are virtually unknown to us now. What is more, we need to consider that with the new possibilities the nanotechnologies are opening up in alliance with biotechnology and medicine—making it possible to treat and potentiating the human body and build organs and tissues for transplantation—there also comes the need to reconsider some core concepts such as moral responsibility. These concepts serve a twofold role: we need them in the first place to ethically and legally qualify these changes and consequences that the potentiating of human capacities will have on future generations, and in the second place to evaluate the limits and the possibilities of their use.

These issues, arising out of the use of nanotechnology and nanomedicine for human enhancement, have been debated from a variety of perspectives—philosophical, ethical, social, and political—and the discussion has come to focus on the following questions (Sparrow 2009):

- How should we address risk?
- How should we regulate medical devices?
- How should we evaluate and deal with the social benefits and harms of emerging technologies?
- What is going to be the impact of these emerging technologies on the health and wellbeing of individuals?

Emerging technologies, genomics, nanotechnology, cognitive enhancement, artificial intelligence, synthetic biology, and biotechnological innovation give rise to controversial situations in research and healthcare practices, as well as national and international law and other domains (see National Nanotechnology Initiative (NNI) http://www.nano.gov/). These controversies involve and influence many human actors (not only citizens, but future generations as well) and this explains the need to focus on managing the consequent risks and properly distributing the associated benefits (see Verdoux 2011; Wolfe 2010). These phenomena also force us to rethink the ethics of responsibility: we need to reorganize the public sphere of rational decision-making so as to take on the question of responsibility freely and collectively. Without certainty about what effects these technological applications will bring about, we should go with the principle of action under which we need to have a predictive view of the future to act in the present. In this paper, I focus on a principle of responsibility that calls for 'prudent vigilance' in assessing benefits next to safety concerns and security risks.[4] I believe that although this principle makes it possible to take into account the

---

[4] In this analysis, I will understand responsibility as equivalent to being responsible and being aware that we can choose how to act and that choices have consequences. On the concept of being responsible, see Haydon (1978).

full range of consequences—both good and bad: benefits and risks—that can be predicted to follow from the use of the emerging technologies in question, it is not so rich as to provide on its own a basis of public policy. I believe it takes more than a balance of technological risks and benefits to arrive at a public and normative ethics for these emerging technologies.

From a legal regulatory perspective, the one distinguishing feature between the EU and other countries is acknowledged to lie in the precautionary principle, which in the Rio Declaration on Environment and Development (1992) reads as follows:

> In order to protect the environment, the precautionary approach shall be widely applied by States according to their capabilities. Where there are threats of serious or irreversible damage, lack of full scientific certainty shall not be used as a reason for postponing cost-effective measures to prevent environmental degradation.

But perhaps the best way to understand this principle is by contrasting it with the prevention principle, under which authorities are required to take action at the earliest possible stage—if at all possible prior to any damage occurring—to prevent known risks from being realized.

Clearly, these two principles take two opposite views of risk analysis, with a focus on *known* risks on the one hand (under the principle of preventive action) and with *uncertain* risks on the other (under the precautionary principle): we proceed on the one hand from the premise that government should only concern itself with those activities that are *certain* to cause environmental damage, and on the other from the opposite premise that no risk can be ruled out with watertight certainty. This analytical distinction can serve to introduce the precautionary principle, but it does not give us a sense of the broader issues in context. Indeed, for one thing, we need to consider that fully-known risks are extraordinarily rare. And, for another, the legal status of the precautionary principle is in dispute. A number of treaties include this principle, and it is certainly a general principle within certain regional contexts, such as the European Union, but it is too early to refer to the principle as a part of public international law.[5]

I believe the two principles just mentioned can only provide a formal abstract framework for an ethics of responsibility, and that they accordingly need to be supplemented with a set of principles operating on the practical level: these practical principles, I submit, should revolve around an idea of the intergenerational moral responsibility, thus serving as a foundation for an ethics of responsibility in matters involving the welfare of people across generations. I further believe it is essential to analyze the foundations of the ethics of responsibility in connection with the due respect for the dignity of the human person and for human rights and fundamental freedoms in the international community. So I am anchoring my analysis to the framework of national and international rules, principles, codes of conduct in bioethics, with specific reference to three documents: (a) the Convention for the Protection of Human Rights and Dignity of the Human Being with regard

---

[5] On the precautionary principle in the context of international law as concerns biotechnology, see Francioni and Scovazzi (2006).

to the Application of Biology and Medicine: Convention on Human Rights and Biomedicine of the Council of Europe, adopted in 1997 and entered into force in 1999, together with its Additional Protocols; (b) the Declaration of Helsinki of the World Medical Association on Ethical Principles for Medical Research Involving Human Subjects, adopted in 1964; and (c) the International Ethical Guidelines for Biomedical Research Involving Human Subjects of the Council for International Organizations of Medical Sciences, adopted in 1982. Also needing to be taken into account is the more recent Universal Declaration on Bioethics and Human Rights (UNESCO 2005), to be understood in a manner consistent with domestic and international law in agreement with human rights law. Article 3, "Human Dignity and Human Rights," emphasizes that human dignity, human rights, and fundamental freedoms are to be fully respected, and that the interests and welfare of the individual should have priority over the sole interest of science or society. Also important is Article 14, "Social Responsibility and Health," under which the progress of science and technology should not undermine the enjoyment of the highest attainable standard of health as one of the fundamental rights of every human being, ensuring as well access to quality health care and essential medicines and to adequate nutrition and water and improved living conditions, a better environment, while also eliminating marginalization and the exclusion of persons on the basis of any grounds.

In constructing an ethics of responsibility that will also take the intergenerational question into account, we need to also consider Article 16, "Protecting Future Generations," under which due regard needs to be given to the impact of the life sciences on future generations, including on their genetic constitution, and we also need to think about the role individuals play in making political decisions now which may turn out to have a big impact on future generations (Usami 2011).

In constructing an ethics of responsibility that will also take the intergenerational question into account, we need to think about the role individuals play in making political decisions now which may turn out to have a big impact on future generations.

I will take responsibility to be a notion whose moral force is grounded in the use of deliberation to frame technology policies by weighing two sorts of considerations, taking into account, on the one hand, what we expect from the emerging technologies and what we want them to do, and, on the other, what the likely outcomes of their use will be and whether or in what respects and to what extent they will still be desirable in light of that projected impact. I start out from the premise that we are morally responsible for all the consequences that our actions have on the wellbeing of future generations, and this suggests the need to deal with emerging technologies on the basis of cooperative strategies through which to evaluate what the best choice is for us and for later generations. To this end I will focus on a set of utilitarian principles, since I believe that utilitarianism offers a basis for deliberating about future choices in a morally responsible way that is neutral and rational with respect to future generations: what is good for us may be bad for people in the future, so we have to take into account the distinction between ends which are good for us, or prudentially good, and those which good for everyone on the whole, or morally good.

As will become clearer in what follows, my method of analysis combines two approaches, for while I frame the issues in light of the principles of responsibility and precaution (starting out from Hans Jonas's conception of them), I ground these principles in a utilitarian foundation. I do so because it seems impossible to me to speak of the use of technologies without considering the *consequences* of such use, and so we cannot avoid a consequentialist (and hence utilitarian) framework of moral evaluation. While I grant that the question of the moral responsibility we have to future generations can be treated from other perspectives (such as contractualism or the theory of rights), I have chosen to do so taking a utilitarian approach because it seems better equipped to solve the problems at hand. Indeed, we are dealing with actions at once collective and social, so on the one hand we cannot evaluate these actions without taking their consequences into account, and on the other we should want to proceed on a deliberative basis in thinking about how to deal with those consequences. As Haydon says 'it is required not merely that one act with regard for the consequences, but that one evaluate and weigh the consequences properly. The agent is expected to pay special attention to consequences involving harm and benefit to others, and to modify his conduct so as to promote benefit or at least to minimize harm' (Haydon 1978).

3.2 Some Models for an Ethics of Responsibility for Future Generations in the Age of Emerging Technologies

As can be appreciated from the foregoing discussion, we are looking at a new problem for ethics to deal with: the problem of the collective responsibility we have to future generations in consequence of our present development and use of emerging technologies, like nanotechnology and nanomedicine. The age of emerging technology puts at the center of ethics the question of the consequences of the actions of *homo faber*: these actions are directed not only at the nonhuman world but also at humans themselves, now more than ever as an object of technology (in medicine, for example, the individual is both the subject and the object of the technology, as can be appreciated by considering how death, formerly regarded as a natural event, is increasingly becoming an artificially controlled process). These new forms of human action call for an ethics of responsibility incorporating a predictive element, because human actions do not only have an immediate effect but are also intended to affect the universal order of things for future generations. And precisely for this reason—that is, because the effects of technology are projected into the future—we can no longer deal with them on the basis of a principle of *subjective* responsibility, which governs private conduct, but must rather look to a principle of *objective* responsibility.

We should bear in mind here that the first Annual Report on Ethical and Social Aspect of Nanotechnology (published on 15 April 2009 by the University of Aarhus, see Malsch and Hvidtfelt Nielsen 2009) discusses responsibility and technology in a dedicated section where we read: 'Nanotechnology must be developed in a safe and responsible manner.' For the reasons I have laid out, I think this concept must be explored from the standpoint of collective responsibility toward future generations. So, from an ethical point of view, the problem is: what kind of moral principle should we rely on in thinking about our present use of technology to make us responsible to future generations? And how should this principle and these considerations inform policy aimed at regulating these emerging technologies? (See Johnson 2007).

The first reference point in dealing with these problems is Hans Jonas's model of responsibility ethics, under which we should act so that the effects of our actions are compatible with the permanence of genuine life (Jonas 1984). Jonas claims that three fundamental assumptions of traditional ethics are in crisis, namely, that (i) the human condition is fixed once and for all, (ii) it is possible to determine what is good for an individual, and (iii) human responsibility can be rigidly defined (Jonas 1974). These assumptions began to fall apart as a result of our enhanced ability to act in the age of technology, and this has opened a new dimension for which there is no precedent in policy and which traditional ethics is ill-equipped to deal with (Jonas 1974). Modern technology, Jonas argues, modifies the human action in a particular way because it makes the individual capable of upsetting the natural order of things and the idea of nature as an unchanging cosmic order. According to traditional ethics, nature is not a direct object of human responsibility and action, and so, the action of the nonhuman is not significant from a moral point of view. Jonas observes that control over events was limited in the past, and the long series of consequences of human action was left to destiny, so traditional ethics focused on the relationship between individuals in private and public life. The wise individual was the one who dealt with the circumstances of life with virtue and wisdom, cultivating these capabilities themselves and leaving everything else to an unknown destiny.

Jonas laid great emphasis on the need to have foresight and consider future scenarios, proscribing all activities that could lead to the extinction of the human race. He took a precautionary approach based on raising awareness of technological risks, and his approach aims to establish limits to the manipulative power of new technologies, but this position gives rise to a paternalistic intergenerational attitude: actions are disallowed wherever there is uncertainty about the future of technological advancements; this is an attitude of prudence (*in dubio pro malo*) toward generations, insofar as they stand to be affected by the consequences of our actions. So, in making choices regarding technologies that are expected to have consequences for future generations, we should not risk harming the interests of these future people, and certainly we should not put their life in jeopardy. The principle of responsibility is based not on the reciprocity of rights and duties but on a primary, nonmutual responsibility. For these reasons, in light of the uncertainty and the irreversibility of some technological choices, Jonas argues for a principle of precaution. So, if prudence is the primary imperative of responsibility, then fear is the preliminary imperative of an ethics of responsibility: it is fear that underlies our taking of responsibility. Is Jonas's approach, proceeding from a 'heuristics of fear,' adequate for deciding about responsible technology?

While I do think we should recognize the importance of Jonas's idea of responsibility for the consequences of our actions, I would also argue that fear and prudence cannot be fashioned into a normative principle: yes, these criteria do offer a way to justify the moral choices we make with respect to future generations, and they make humanity (human identity and nature) a frame of reference in thinking about human responsibility and free action, but they cannot form a basis on which to reason about the values we should strive to realize through action, especially when grappling with new situations like the ones brought about by the new technologies. Jonas's conception, then, is only useful as a starting point in searching for an ethics of responsibility suited to the kinds of evaluations we need to make in dealing with the emerging technologies.

So, while I recognize the appeal of Jonas's (1984) principle of responsibility—requiring us to act compatibly with the dignity of human life in the distant future—for it rightly places the human condition at the centre of our practical reasoning about what to do, I do feel that this principle lacks moral force (because it is grounded on the nonmoral notions of fear and prudence) and it falls short of giving us a full conception, for it is based on a loose understanding of human life. We also need to take into account what it is that makes us *specifically* human, and here we cannot discount the idea of humans as autonomous individuals capable of making their own evaluations about what the best practical choices are: in this way we have a basis on which to say (a) what accounts for the dignity of human life, and (b) why we should be responsible for preserving human life as an inherent good. It is along these lines that Engelhardt reasons in setting out the idea of the morality of wellbeing, arguing that our responsibility to future generations is not prudential but moral: we are morally autonomous beings, and as such we are individually responsible not only for our own psychological and physical wellbeing (or quality of life) but for that of others, now and in the future.

From this perspective—that of Engelhardt's morality of wellbeing—we do not take a paternalistic attitude toward humanity and technological advancements: everybody is free to choose and decide (Engelhardt 1996). And so action requires only that we preserve each individual's right to self-determination and that we provide the conditions enabling that right to be exercised, which in turn means that we must relate to one another on the basis of a moral principle of tolerance. From this conception also flows a principle of prudence—we must act in such a way that costs do not outweigh benefits—and in this respect Engelhardt's conception is similar to Jonas's, but unlike Jonas, Engelhardt grounds this principle in a distinctly moral conception of the person from which flows his chain of reasoning: we are all autonomous individuals; this endows each of us with a right to self-determination; and so we must secure the minimum conditions enabling us to each exercise this right in peaceful coexistence—hence the need for prudence and tolerance, especially in view of the plurality of moral perspectives that individual self-determination gives rise to. These plural perspectives compete for control of policy-making, and the moral framework within which we reason in making such choices must therefore be capable of holding together a pluralistic society like ours. This in turn means that moral authority is not to be found in any one value system but rather derives from the consent of those who interact and participate in shaping a world that everyone is going to be affected by. We each live in one or more thick moral communities, and although we can choose to enter or leave these communities at will—espousing or rejecting their belief systems and their systems of morality—we cannot choose to exit the larger society: it follows that if we are to have a life in different communities bound by different systems of shared moral content, all the while enabling these communities and their members to form a society, we must frame society without reference to any thick moral content beyond the content necessary to enable the previously mentioned peaceful and functional coexistence.

The two models of ethics so far discussed, Jonas's and Engelhardt's, have been designed by the authors in relation to technological advances in genetic engineering. If we transfer these models to the techno-scientific expectations about other emerging technologies, such as nanotechnology, the results will be as follows: Jonas would focus on the question of the limits to be placed on the individual's manipulative power, while Engelhardt would insist on the legality and positivity of human intervention on living matter. Jonas would say that we have to close the Pandora's box of enhancement technologies because it is irresponsible to develop new potentially risky technologies for future generations. Engelhardt would say that any kind of technological innovation is acceptable, since it will give us the power to be the designers of nature, giving us a chance to further enhance and develop it.

Recently, Habermas has focused on the acceptability of genetically engineering human beings. Habermas rejects the notion that we should improve the natural endowment of fetuses and adults, and claims that genetic improvement raises complex moral questions in both cases, for we should want to protect the right to a nonmanipulated genetic identity (Habermas 2003). We cannot independently decide for ourselves against intervening on one's own future genetic heritage, and this impossibility to decide leads to asymmetrical relations of power and responsibilities between generations: no unborn person can consent to genetic treatment, and future individuals will not be fully free if they are subject to genetic programming.

Habermas argues that if we allow such improvements, we risk changing what it is that makes us human (our human nature), and that relations among individuals would be dictated by the exchange between what is technically a 'product' and what is 'natural' (Habermas 2003). In this sense, and arguing from the premise that we need to protect the biological basis of the individual, Habermas aims to defend the capacity for self-determination and mutual respect of human beings and their self-understanding.

These models highlight some critical factors that figure in the shaping of the imperative of responsibility: this imperative brings up the issue of the moral collective responsibility for emerging technologies, and it forces us to take into account our long-term interests and the need for socially and environmentally responsible behaviours.

Habermas and many other moral philosophers have dealt in different ways with the notion of collective responsibility for future generations. Two of these philosophers are Sandel and Harris, who both try to capture the moral and social challenge we face with scientific and technological development, namely, the consequences of our use of technology and the implications of our collective actions. While these moral theories give us different useful perspectives on the concept of responsibility for future generations, they fail to resolve the thorniest of the questions at issue: how are we to conceive the autonomy, preferences, and interests of future generations once we realize we can use nanotechnology and neuroscience to enhance or modify existing capacities and to restore abilities that may have been lost?[6]

This shortfall suggests that we look to a different approach to intergenerational moral responsibility, an approach that has us evaluate each specific situation by balancing risks and opportunities.[7]

One strategy in this sense would consist in (a) gathering all the data and making our best predictions about the likelihood that our actions today may have adverse consequences in the future, and (b) making on that basis a value judgment as to whether those actions are advisable. And on Jonas's ethics of responsibility, we would prohibit any actions liable to trigger a process whose exact consequences are unknown. But we have to consider that our moral collective responsibility cannot be based solely on the likelihood of negative risks. Caution is important in many bioethical decisions based on the principle of responsibility, but it cannot alone guide the development of such technologies. From an intergenerational moral perspective, we ought to avoid predicating our moral collective responsibility on a paternalistic view based on the precautionary principle, precisely because, as Habermas notes, we cannot decide for others what will be good for them. So it is a different set of considerations that we should take into account in shaping the principle of responsibility as a moral criterion:

- We should consider the consequences of technologies in terms of empirical practices and worry about future generations, doing everything possible to avoid damaging and limiting individuals' freedom.
- We should create the conditions that will maximize their freedom and minimize any damage to them.
- We have a moral duty to promote research and experimentation geared toward human improvement, but we must also ensure that this does not lead to bad irreversible effects.

3.3   The Utilitarian Proposal

It can be appreciated from the discussion just ended that technological development needs to be dealt with through an ethical framework that takes into account not only individual freedom and autonomy but also other people and the whole of society. We are currently witnessing a rapid development of technologies: advances in genetics, nanotechnology, and neurotechnologies are but a few examples. The most developed of these are perhaps genetic technologies, with the routine use of genetic testing of patients,

---

[6] On these approaches, see Sandel (2007) and Harris (2007). See also Savulescu and Bostrom (2009).

[7] On this point see Lucivero et al. (2011).

in vitro fertilization, and DNA techniques. But the same sort of development can be observed in nanotechnologies: nanoproducts are already available on the market, and others are in the pipeline. Some researchers expect that in the future some medical testing will be done through ingested nanobiosensors that can detect items such as blood type, DNA, and drugs. Neurotechnologies are similarly evolving, and we can already use various brain-scanning devices and pharmaceutical-treatment techniques.

These advancements make it all the more urgent to decide on a policy level whether we should regulate human-enhancement technologies by preventing or mitigating the negative effects and fostering the positive ones.

Yet it is important to keep in mind that the human-enhancement debate is not just a theoretical discussion about ethics. This debate has a bearing on real-world and policy decisions and can affect not just those who would be enhanced but also researchers and social institutions, as well as our ideals of freedom and human dignity. Although these technologies are not fully developed, it is not unreasonable to expect that they will keep marching along their groundbreaking path and bringing with them an increasing cluster of new ethical issues. These technologies, and nano-technology in particular, have an essential revolutionary feature: this is their material malleability, a feature by virtue of which our lives can be improved in both quality and quantity. Clearly, this would mean a significant impact on society.

The material malleability of nanotechnology means that ingredients can be arranged so as to self-assemble, in such a way that they can be created in large quantities. Some researchers suggest that nanostructures could assemble other nanostructures or could self-replicate. How all of this is possible remains an open empirical question. In any event, nanotechnologies will significantly impact society. If minds are brains and neurotechnology will enable us to construct and manipulate them, these technologies will be more revolutionary than any other technology so far. Minds could be altered, improved, and extended in ways we cannot even imagine.

The regulation of these activities is still an unsettled matter, especially as we are only beginning to understand how the issues ought to be framed, how we should go about addressing them, and what is at stake. Now, but also *in* and *for* the future, we have to decide whether we should have (i) no restrictions on the development and use of these technologies, or (ii) some restrictions, or (ii) a full ban.

These decisions we cannot make without discussing the operative concept of moral responsibility to future generations. In what follows I will outline a model of ethics that could be useful in thinking about individual, collective, and institutional choices that can be expected to significantly impact future generations. The two models previously discussed have shown us how to deal with empirical questions when it comes to deciding on the wellbeing of future generations, but neither of them offer operative principles we could rely on in defining moral responsibility. For this reason, I think it is more useful to look to a dynamic approach, one based on a set of principles that can be used in framing a cooperative strategy. As I previously suggested, these principles can be arrived at working from a utilitarian perspective.[8] One of the utilitarian principles that could support us in making future

---

[8] On the rational and moral foundation of a utilitarian ethics, see Gandjour (2007).

choices requires us to act in such a way as to maximize total utility, understood as the greatest possible amount of guarantees. This is a quantifiable method of deliberation in the sense that it identifies a just decision as the option yielding the highest expected utility, which is computed by considering what the consequences of a certain course of action would be, putting a positive or negative value on each of those consequences, scaling those values according to the likelihood that the relative consequences will actually materialize, and finally subtracting the negative values from the positive ones (the algebraic sum of these values will yield the expected utility of pursuing a given course of action). However, there are situations where it will prove problematic to apply this method of deliberation, because we cannot always estimate the probabilities of the various outcomes of different alternatives. In fact, in some situations this method will even be irrational, leading to catastrophic consequences: this is so when one of the alternatives may result in irreversible effects. This is a problem one could solve by relying on the maximin principle, under which we must only look at the worst outcome of each alternative and then choose that alternative whose worst outcome is better than the worst outcome of any other. But even this idea presents difficulties: it leads to paradoxes because it implies that each alternative is disqualified as the worst. Where neither the principle requiring us to maximize expected utility nor the maximin principle can be applied, we could use other methods that involve a combination of seeking best possible outcome while avoiding the worst. It appears rational to reject an intervention if there is some risk of its causing a grave enough harm, even if its expected utility is mathematically greater than its expected disutility (this is not true in all cases, and even if the expected utility of an action outweighs its expected disutility, there may be important reasons not to pursue it despite that positive balance).

But what needs to be said is that a number of factors make it difficult for us to embrace a specific responsibility toward future generations. One of these factors is that we are expected to be able to correctly predict the preferences that individuals in the future will have in their decision-making processes. In addition, the choices we *currently* make involve forms of cooperation among countless people who do not collaborate with one another but rather act according to their own personal interest, and this eventuates in a collective loss (such as the greenhouse effect and other forms of environmental damage). And, finally, there is the free-rider problem, i.e. the risk that some may act selfishly by exploiting collective resources which everyone contributes to but which the individual in question (the free rider) does not (in effect riding on everyone else's back).

But the point of the utilitarian perspective is to work out moral criteria on which basis to achieve an operative moral intergenerational responsibility: these criteria are designed to guarantee the vital interests of future generations, but without reducing the options of those in existing generations. These moral criteria are then translated into policy measures, which consist in enacting national or international laws and in creating institutions capable of enforcing these laws. If we want to build philosophical and ethical operating principles, we should have policies that instead of just pursuing each state's national interests also take in a global perspective, since the problem of responsibility toward future generations is not limited by national boundaries. In this sense, one formal requirement based on utilitarianism (and which could also be included in the normative model for regulating future uses and applications of nanotechnologies) is to implement forms of 'contributory public responsibility,' that is, forms of responsibility where responsibility is ascribed not only to public organizations—national and international government agencies entrusted with framing policies on the use of emerging technologies—but also to the individuals who have had an active role in designing those organizations and framing those policies. This suggests a way to frame the deliberative public process so as to directly and effectively involve the representatives of multiple social communities (Sparrow 2009). These representatives should express the needs and concerns of their communities, put forward strategies for addressing needs and concerns, and assume responsibility for the consequences these strategies may lead to. And at the same time they should be open to taking on broader, normative responsibilities as the need for them becomes apparent: these are the responsibilities which come with the possibilities afforded by nanomedicine and the nanotechnologies, making it possible to modify the environment, lifestyles, and capacities of humans in future generations.

3.4 Conclusion

I do not believe we need any *new* concept of moral responsibility for dealing with emerging technologies: what we need to do is not devise new concepts in nano-ethics, neuro-ethics, or gene-ethics, but to reframe the concepts we already have in light of the processes of technological development (Moor 2005), and in this way we can already hope to have useful guidance on the course we should take moving forward. This guidance, I think, comes to us by way of a utilitarian conception, yielding dynamic operative principles on which basis to act in dealing with the process of technological development.

There is an old Indian parable that may be worth recalling here, for it can be used to point out how a specific new ethics of responsibility for future generations can be myopic. In this parable, six blind men come across an elephant but none of them know what an elephant is. Thus they move close to the animal so as to feel it and try to get a sense of what it is. Thereafter they relate their experiences, but since they each felt a different part of the elephant, one man describes the animal as a pillar (he had felt the leg), another one as a fan (he had felt the ear), another as a wall (the torso), another as a rope (the tail), and so on. Each man experiences only a part of the elephant, but each insists he is correct in describing the *whole* elephant. The problem, of course, is that none of the blind men could see the *entire* elephant, and so they cannot understand what an elephant really is.

I submit that we run the same sort of risk with the overlapping ethical issues arising in connection with emerging technologies (nanotechnologies, genetics, the neurosciences, neurotechnologies) and other scientific and technological domains that give rise to such issues.

Once technologies enter a revolutionary stage, their social impact inevitably gives rise to moral issues of the sort we are concerned with. This phenomenon happens not simply because an increasing number of people are affected by the technologies but also because the technologies inevitably provide numerous novel opportunities for action for which we have no moral compass, and hence no policy framework within which to deal with the consequences of those new opportunities. Such moral issues force us not only to think about new human values and interests but also to rethink some basic ones, such as autonomy, the quality of life, identity, privacy, and moral responsibility. These moral issues also require considerable effort before we can formulate and justify good policies for them. Surely, we have to improve our ethical approach to technology. We need to understand that ethics is an ongoing and dynamic tool: we cannot anticipate every moral issue that will arise from a developing technology. Because of the limitations of human cognition, our moral understanding of developing technologies will never be complete. We have to do as much as we can while taking into account that applied ethics is a dynamic tool that calls on us to constantly reassess the situation. What could improve the moral debate is a better collaboration among ethicists, scientists, technologists, and social scientists. We need a multidisciplinary approach: public forums need to be created where scientists can lay out future developments, citizens can have open debates about them, decisions-makers can be entrusted with developing policy guidelines and regulations for these new technologies, and everyone is collectively responsible throughout the process.

Ethicists need to be informed about the nature of technology and need to proceed on an empirical basis to investigate its nature and consequences. Scientists and technologists need to take into account the considerations raised by ethicists and social scientists.

The ethics of technology, bioethics, the ethics of medicine, and even the philosophy of technology concern themselves with questions of sustainability, risk assessment, and the intersection between human beings and technology. The perils and benefits of emerging technologies make it necessary to press on with a research programme aimed at understanding the underpinnings of morally responsible behaviour.

References

Chadwick, R. 2008. Therapy, enhancement and improvement. In *Medical enhancement and post-humanity*, ed. B. Gordijn and R. Chadwick, 25–37. Dordrecht: Springer.

Council of Europe. 1997. *Convention for the protection of human rights and dignity of the human being with regard to the application of biology and medicine: Convention on human rights and biomedicine.*

Declaration of Helsinki. 1964. *World Medical Association on ethical principles for medical research involving human subjects.* Adopted by the 18th WMA General Assembly, Helsinki, Finland, June 1964; amended by the 29th WMA General Assembly, Tokyo, Japan, October 1975; 35th WMA General Assembly, Venice, Italy, October 1983; 41st WMA General Assembly, Hong Kong, September 1989; 48th WMA General Assembly, Somerset West, Republic of South Africa, October 1996, and the 52nd WMA General Assembly, Edinburgh, Scotland, October 2000. http://www.who.int/bulletin/archives/79(4)373.pdf

Engelhardt Jr., H.T. 1996. *The foundations of bioethics*, 2nd ed. New York: Oxford University Press.

Francioni, F., and T. Scovazzi (eds.). 2006. *Biotechnology and international law*. Oxford/Portland: Hart Publishing.

Gandjour, A. 2007. Is it rational to pursue utilitarianism? *Ethical Perspective: Journal of the European Ethics Network* 14(2): 139–158.

Habermas, J. 2003. *The future of human nature*. Cambridge: Polity Press.

Harris, J. 2007. *Enhancing evolution: The ethical case for making better people*. Princeton: Princeton University Press.

Haydon, G. 1978. On being responsible. *The Philosophical Quarterly* 28(110): 46–57.

International Ethical Guidelines for Biomedical Research Involving Human Subjects. 1982. *Council for International Organizations of Medical Sciences*. (CIOMS) http://www.cioms.ch/. (First published in 1982, revised in 1993 and finally in 2002)

Johnson, D.G. 2007. Ethics and technology 'in the making': An essay on the challenge of nanoethics. *Neuroethics* 1(1): 21–30.

Jonas, H. 1974. *Philosophical essays: From ancient creed to technological man*. Chicago: Chicago University Press.

Jonas, H. 1984. *The imperative of responsibility: In search of an ethics for the technological age*. Chicago: The University of Chicago Press.

Levy, N. 2011. Enhancing authenticity. *Journal of Applied Philosophy* 28(3): 308–318.

Lindsay, R.A. 2005. Enhancements and justice: Problems in determining the requirements of justice in a genetically transformed society. *Kennedy Institute of Ethics Journal* 15(1): 3–38.

Lucivero, F., T. Swierstra, and M. Boenink. 2011. Assessing expectations: Towards a toolbox for an ethics of emerging technologies. *Neuroethics* 5(2): 129–141.

Malsch, I., and K. Hvidtfelt Nielsen. 2009. *Individual and collective responsibility for nanotechnology*. First annual report on ethical and social aspects of nanotechnology by the ObservatoryNano project.

Moor, J.H. 2005. Why we need better ethics for emerging technology. *Ethics and Information Technology* 7(3): 111–119.

National Nanotechnology Initiative (NNI). n.d. *Nanotechnology fact*. http://www.nano.gov/publications-resources. Accessed 20 Dec 2012.

Orlebeke Caldera, E. 2008. Cognitive enhancement and theories of justice: Contemplating the malleability of nature and self. *Journal of Evolution and Technology* 18(1): 116–123.

Rawls, J. 1971. *A theory of justice*. Cambridge, MA: Harvard University Press.

Rio Declaration on Environment and Development. 1992. *United Nations Conference on Environment and Development (UNCED)*. http://www.un.org/documents/ga/conf151/aconf15126-1annex1.htm. Accessed 24 Jun 2014.

Sandel, M. 2007. *The case against perfection: Ethics in the age of genetic engineering*. Cambridge, MA: Harvard University Press.

Savulescu, J., and N. Bostrom (eds.). 2009. *Human enhancement*. Oxford: Oxford University Press.

Savulescu, J., and I. Persson. 2008. The perils of cognitive enhancement and the urgent imperative to enhance the moral character of humanity. *Journal of Applied Philosophy* 25(3): 162–177.

Schermer, M. 2011. Health, happiness and human enhancement—Dealing with unexpected effects of deep brain stimulation. *Neuroethics* 6: 435–445. doi:10.1007/s12152-011-9097-5.

Sparrow, R. 2009. The social impacts of nanotechnology: An ethical and political analysis. *Journal of Bioethical Inquiry* 6(1): 13–23.

UNESCO—United Nations Educational, Scientific, and Cultural Organization. 2005. *Universal declaration on bioethics and human rights*. Available from: http://www.unesco.org/new/en/social-and-human-sciences/themes/bioethics/bioethics-and-human-rights/. Accessed 31 Jan 2014.

Usami, M. 2011. Intergenerational rights: A philosophical examination. In *An anthology of philosophical studies*, vol. 5, ed. Patricia Hanna. Athens: Athens Institute of Education and Research.

Verdoux, P. 2011. Emerging technologies and the future of philosophy. *Metaphilosophy* 42(5): 682–707.

Wolfe, C. 2010. *What is posthumanism?* Minneapolis: University of Minnesota Press.

Chapter 4
The Role of Responsible Stewardship in Nanotechnology and Synthetic Biology

Ilaria Anna Colussi

4.1 Introduction

In these current times, scientific and technological progress is a global phenomenon. Theoretical discoveries in the fields of medicine, life sciences, neurosciences, cognitive science, and practical applications derived from them constitute two interconnected and complementary dimensions that create a rich and complex scenery. Furthermore, a trend of interaction and convergence between different sciences and technologies is visible. A meaningful example of this is represented by 'nanoscience' and 'nanotechnologies' that – like the other emerging technologies – demonstrate numerous beneficial applications, thus entailing potential benefits, but at the same time pose certain risks and concerns.

This paper focuses on some of the risks that have arisen in the context of nanotechnologies, in order to look for and articulate a model of governance for dealing with these risks. For achieving this aim, a methodology founded on the comparison with another 'converging science'–the so-called synthetic biology (henceforth 'synbio')–has been chosen.

From the structural viewpoint, the proposed analysis will be developed as follows: (a) a framework of similarities and differences between synbio and nanotechnology is offered; (b) an examination of the risks and the approaches required to deal with them is given, by means of a critical study of the different models of governance, and (c) this analysis proposes to find, in the end, the most suitable and proper solution to apply in the area.

As a premise, it should be noted that the focus of this paper is on health, safety and security issues. The social, political, economic and more intrinsically ethical topics, which are commonly associated with nanotechnologies and synbio, will not

I.A. Colussi (✉)
Doctoral School of Comparative and European Legal Studies,
University of Trento, Trento, Italy
e-mail: ilariaanna.colussi@gmail.com

S. Arnaldi et al. (eds.), *Responsibility in Nanotechnology Development*, The International
Library of Ethics, Law and Technology 13, DOI 10.1007/978-94-017-9103-8_4,
© Springer Science+Business Media Dordrecht 2014

be taken into consideration. Thus, the 'risks and benefits' framework that is the point of reference in this paper refers only to those implications of nanotechnologies that can affect human health, environment, and public security.

4.2 Definitions

"In the beginning, there was the Word" ("Εν ἀρχῇ ἦν ὁ Λόγος").[1] First of all, it is necessary to start from the definitions of synthetic biology and 'nanoscience/nanotechnology'.

The expression 'synthetic biology' was first used in 1974 by Waclaw Szybalski, who saw the potential of molecular biology to evolve from the description to the manipulation of genetic systems (Szybalski 1974). In 1980, the term 'synthetic biology' was chosen by Barbara Hobom to describe bacteria that had been genetically engineered using recombinant DNA technology (Hobom 1980). In this sense, 'synthetic biology' was synonymous with 'bioengineering'. In 2000, it was used to describe the synthesis of unnatural organic molecules that function in living systems, by some speakers at the annual meeting of the American Chemical Society in San Francisco (Rawls 2000).

Over the years, the meaning of synthetic biology has been adopted to express the core concept of (1) redesigning life (Szostak et al. 2001), through 'the re-engineering of existing biological elements' (High-level Expert Group 2005, p. 1) and (2) designing and fabricating novel ('synthetic') biological components, meant as the engineering of biological components and systems that do not exist in nature. In other words, synthetic biology is a discipline of the intersection between engineering and biology, but at the same time it is also a synergistic 'new trend in science and technology and a clear example of converging technologies, i.e. nanotechnology, biotechnology, information technology and cognitive sciences' (De Vriend 2006, p. 9; Calladine and Meulen 2012).

As for the terms 'nanoscience and nanotechnology', it appears that they too arise from the convergence of different aspects of scientific research. Indeed, nanotechnology is not a separate techno-scientific field. Rather, it is a new platform for a range of existing disciplines–including chemistry, physics, biology, biotechnology, neurology, information technology and engineering – which allows a shift down to the nano scale[2] (ETC Group 2003). Nanoscience is intended as 'the scientific study, on the atomic and molecular scale, of molecular structures one of whose dimensions measures between 1 and 100 nm, with a view to understanding their particular physico-chemical properties and to defining the means required to manufacture, manipulate and control them' (Gouvernement du Québec 2006, p. i). Thus, nanotechnology flows from nanoscience and consists of 'the design, characterisation, production and application of structures, devices and systems by controlling shape and size at the nanometre scale' (The Royal Society and The Royal Academy of Engineering 2004, p. 5).

---

[1] So begins St. John's Gospel.

[2] As we know, one 'nano' is one billionth of a meter or 10^{-9}.

The birth of the nano field is officially attributed to Richard Feynman's pioneering speech on atomic engineering, 'There is plenty of room at the bottom' (Feynman 1959), but the term 'nanotechnology' was coined by Norio Tanighuchi and popularised by Eric Drexler in the book *Engines of Creation* (Drexler 1986), where the author imagined nano-scale machines operating with atomic precision and made by hard materials like diamond to fabricate complex nano-scale structures.

In a nutshell, both synthetic biology and nanotechnology are converging technologies. However, on the one hand, the notion of 'synbio' seems to be broader as it includes nanotechnology as well and 'embraces the biological strands of nanotechnology' (Ball 2005, p. R1). On the other hand, synthetic biology is part the fourth generation of nanotechnology applications, thus making it an evolution of nanotechnology, which should be subsumed within it (Roco 2006).

4.3   The Comparison Between Nanotechnologies and Synthetic Biology

The comparative analysis between nanotechnologies and synbio is now developed through the focus on technical/scientific aspects, approaches, applications and concerns that have arisen in the context of the two emerging technologies.

4.3.1   Materials and Products

Till the present, numerous classifications of synthetic products have been suggested. Bhutkar (2005) proposes to subdivide synthetic products into the four following categories: (a) *synthetic elements* (the fundamental building blocks providing primitive functionality); (b) *synthetic networks* (composed of interacting components); (c) *synthetic organisms* (the result of synthetic assembly of complete or minimal genomes of an organism); and (d) *synthetic systems* (representing the ultimate goal of synthetic biology, i.e. the aim to design synthetic systems composed of multiple synthetic organisms).

In the field of nanotechnology, experts usually work with *nanomaterials* (intended as chemical and biological compounds), using them to produce nanoparticles by breaking down those elements into nano-scale bits, as well as to manufacture distinctly new materials (Maynard 2006). The main materials are *nanoparticles,* subdivided into: (a) *synthetic nanoparticles* (i.e., engineered or manufactured nanoparticles, which have either two dimensions at the nanoscale level, namely the nanotubes, or three dimensions, such as quantum dots, metallic oxides, carbon black, metals such as gold and silver, semiconductors and so forth); (b) nanoparticles, which are produced by natural combustion processes, such as forest fires; (c) nanoparticles that are the unintentional by-products of human-induced combustion processes, such as cigarette smoking (The Royal Society and The Royal Academy of Engineering 2004). The products of nanotechnology are *nanostructures (aggregate and agglomerate) and nanosystems.*

Thus, nanotechnology and synbio are similar with regards to the components they are dealing with. More precisely, synthetic biology seems to include nanotechnology, by aiming to build new forms of life with elements that have been engineered-in at a nanometer scale.

4.3.2   Approaches

Another point of convergence among the disciplines is given by their approaches: even if there are plenty of approaches followed in the field of synthetic biology[3] as well as in the nanotechnology sector,[4] particular attention has to be given to the top-down approach (de-construction) and the bottom-up approach (construction).

The top-down approach in nanotechnology is the reduction of structures into nano dimensions (through lithography techniques used, e.g., in electronics to produce silicon chips). In synbio, this approach is the cutting and sculpting, i.e., the redesign of the existing organisms or gene sequences with the goal of stripping out unnecessary parts, or replacing or adding specific parts to achieve new or amplified characteristics and functions (e.g. the case of 'minimal genome organism', Glass et al. 2006).

On the other hand, the bottom up approach in nanotechnology is the assembly of molecules like 'building blocks', in order to create nanostructures (such as the manufacturing of carbon nanotubes through chemical synthesis processes, self-assembly/self-organization/positional assembly processes). In synbio, this approach is the design of new biological parts and systems inspired by general biological principles, using biological or chemical components (the so called 'biobricks') and assembling them like Lego pieces, in order to reproduce the behaviour of living systems (e.g. the creation of the first living and replicating bacterium called 'Synthia', having a completely synthetic genome, Gibson et al. 2010).

4.3.3   Potential Applications

The possible applications of synbio and nanotechnologies are recognisable in the following areas:

- *environment:* in the field of bioremediation synthetic microorganisms could be used to degrade pesticides (De Lorenzo 2008), and to minimize water use and

---

[3] See, for example, Deplazes (2009), who proposes a subdivision in bioingeneering, synthetic genomics, protocell synthetic biology, unnatural molecular biology, in-silico approach; O'Malley et al. (2008), who suggests only a division in DNA-based device construction, genome-driven cell engineering and protocell creation.

[4] Beyond top-down and bottom-up approaches, there are: the functional approach (seeking to develop components of a desired functionality without regard to how they might be assembled), and the biomimetic approach (applying biological methods and systems found in nature to the study and design of engineering systems and modern technology). See Boncheva and Whitesides (2004).

replace chemical fertilizers (Kirby 2010). Nanostructured traps are implemented for removing pollutants from industrial effluents and for recycling materials (US National Science and Technology Council 1999);

- *agriculture and food:* synthetic products could be created in order to gain nutritional benefits or to function as biosensors, which monitor the soil for its nutrient quality (Synthetic and Genomics Inc. 2009). The production of monitoring and dosing nanosystems should also improve plant production and food processing, helping to reduce the amount of chemicals used (Scrinis and Lyons 2010);

- *industrial and energy applications*: synthetic biology opens the door to the production of '*biofuels*', which are intended as carbon-neutral (or more environmentally friendly) sources of energy and renewable energy sources derived from biomass (Savage et al. 2008). Nanotechnology helps creating new types of batteries, producing artificial photosynthesis for clean energy, safe storage of hydrogen, energy savings from using lighter materials and smaller circuits. Furthermore, nanotechnology is used within automotive, aeronautic, electronic, information technology and communication industries;

- *health applications*: synthetic products could be used in order to produce medicines (biopharmaceuticals) and vaccines (Mast and Ward 2008; Ro et al. 2006). There are already *in vivo* applications, i.e. designed regulatory circuits to trigger insulin production in diabetes or bacteria/viruses programmed to identify malignant cancer cells and deliver therapeutic agents, or to be used for tissue repair/regeneration, or as vectors for therapy (Serrano 2007). Nanotechnology is employed to produce structured drugs, drug delivery systems targeted to specific sites in the human body, biocompatible replacements for human body parts and fluids, self-diagnostic tests for home use, sensors for labs-on-a-chip and cancer detection, materials for bone and tissue regeneration.

4.3.4   Risks

Synbio and nanotechnologies pose the same risks (Organization for Economic Cooperation and Development (OECD) 2003). The focus of this paper is to examine the risks towards environment, human health, public security, setting aside social, economic, ethical problems that the two emerging technologies generate (such as in the field of intellectual property rights, access to resources, distributive justice, enhancement, improvement of humans, privacy, manipulating life, 'playing god', etc.). However, the concerns about safety and security do have ethical implications. The role of ethics emerges while considering the necessity of selecting the best approach for governing the situation of uncertainty, which is an integral part of the notion of 'risks'. Indeed, being in a situation of uncertainty, 'the contribution of ethics to this subject [...] lies in a value judgement of the situation (relationship of reliable knowledge to the degree of uncertainty)' (Grunwald 2005, p. 187).

4.3.4.1   Bio- and Nano-safety (Health and Environmental Risks)

The World Health Organisation (WHO) defines biosafety as 'the containment principles, technologies and practices that are implemented to prevent the unintentional exposure to pathogens and toxins, or their accidental release' (WHO 2006, p. 7). In general, the main feature of this type of risk is its accidental nature, and not the result of man's ill intentions (i.e., terrorism).

In terms of health applications, the perils of synthetic biology are linked to the unknown consequences that it can have. Infectious diseases may be accidentally transmitted to laboratory workers or to family members following airborne transmission of disease agents, manipulated using synthetic biology techniques. In patients, the use of cell therapies of bacterial microbial origin may cause infections or unexpected immune responses (Schmidt 2009). Similarly for nanotechnologies, the same problems could occur, since nanoparticles have great mobility and almost unlimited access to the body (entering the bloodstream through breathing and via the digestive tract, thus penetrating all the organs). Despite it being unclear about the transmission process, possible ways of transmission could be via damaged skin or through the lymphatic system and lymph nodes (Swiss Re 2004).

As for the environment, there is the risk of harm to plants or animals from the accidental release of synthetic organisms that are difficult to control. These synthetic organisms can replicate uncontrollably, and thus increase pesticide resistance and the growth of new invasive species, all of which have detrimental effects on ecosystems and biodiversity (Eggers et al. 2009). As for nanomaterials and synthetic unbound particles that are insoluble and not biologically degradable, they might not dissolve or degrade without setting off a toxic reaction, and thus they constitute an entirely new class of pollutants.

4.3.4.2   Bio- and Nano-security (Public Security Risks)

This type of risk refers to the misuse and the mishandling of synthetic and nano products and its constituent knowledge by unauthorized people. In this case, biosecurity must be understood in a laboratory context, as the 'control and accountability for valuable biological materials [...] within laboratories, in order to prevent their unauthorized access, loss, theft, misuse, diversion or intentional release' (WHO 2006, p. 7), i.e., all the measures and efforts that are taken and needed to prevent the creation of deadly pathogens for the purposes of terrorism.

In the area of synbio, researchers have demonstrated that it is possible to create or recreate deadly viruses such as polio (Cello et al. 2002) and the 1918 Spanish flu (Tumpey et al. 2005). So, the hypothesis that someone could spread the organisms all over the world to contaminate the environment and to harm humans' health is not so abstract after all.

Linked to the misuse and mishandling of synthetic products for terroristic purposes, there are the dangerous characters such as the 'lone wolf', who is a highly trained biologist, a professional researcher that has access to lab equipment or is working alone as 'garage biologist', and the 'biohacker' who tries to create a virus 'out of curiosity or to show his technical prowess' (Whittall 2009, p. 27). In addition, we should not forget the current birth of movements that are away from the academic lab (such as the Do-it-yourself movement), aiming at bringing biology into the houses and to the masses, or the iGEM competition (International Genetically-Engineered Machine), established in 2005 for gathering students and lecturers from universities across the world, in order to work with biobricks (i.e. standardized and registered DNA elements) for engineering new metabolic pathways in bacteria or eukaryotic cells. All these projects stimulate a new vision of biology as available and manageable for everyone. So, it is apparent that these movements can promote innovation, but at the same time they exacerbate the risk of misuse of the technology (Dietrich and Steen 2007).

As for nanotechnologies, the risk manifests itself when nanoparticles are modified and manipulated in order to kill or to harm people, by exploiting the reactivity developed by certain nanometric particles. In particular, metallic nanopowders may generate the risks of explosion, flammability or toxicity, or they may penetrate the body.

In summary, both synbio and nanotechnologies give rise to the so-called 'dual-use dilemma' (UK Parliamentary Office of Science and Technology 2009), i.e., the dilemma arising when scientific knowledge could be used in both good and harmful ways.

4.4   How to Deal with Risks?

After establishing the basic similarities and differences between nanotech and synbio, the subsequent focus of this paper is on the treatment of risks.

In general terms, our society is based on risks (Beck 1992).

When we deal with new technologies, such risks appear to become bigger and uncontrollable, since their consequences are often not entirely known, difficult to predict, and potentially having catastrophic effects. For this reason, nanotechnologies and synthetic biology could be labelled as 'inchoate technologies', because of 'their ability to evolve in unpredictable ways and to spawn new chains of technological developments' (Gervais 2010, p. 669).

The necessity for the governance of the risks associated with these new technologies requires the adoption of different models. In this paper, the analysis will concentrate on the 'risk analysis' pattern, which is the traditional model for any kind or risk (not only in the context of new technologies, but in the bank, industrial, environmental fields). This pattern is based on three phases: 'risk assessment, risk management and risk communication' (Aven 2008).

In 'risk assessment', the scientific element emerges in the identification of potential risks that arise from a determinate technology. It seeks to evaluate its risk level according to quantitative data or based on perceptions or on economic elements or on trade-offs. It also considers the probability of when such risks could occur (The Royal Society 1992). According to the 2000 European Commission Communication and the 1993 US National Research Council, the phase of risk assessment is subdivided into: hazard identification, dose-response assessment, exposure assessment and risk characterization.

'Risk management' is the phase of evaluation, policy and decision, where possible actions for regulating a new technology ('policy phase'), at the light of the mentioned analysis of risks, are weighed and considered.

The final phase of 'risk communication' consists of making the public and the various stakeholders become aware of the chosen policy in a transparent manner. It is a duty that is generally adopted by the mass media, which has the power to influence public opinion, trust, the acceptability or the refusal of a new technology (Slovic 2000).

The described model above is interpreted in several ways, in accordance to the principles that intervene in the phase of risk management as policy tools.[5] The main principles considered here are (a) the 'precautionary principle', according to which a technology should be considered dangerous until proved to be safe, in line with the commonsense motto 'better safe than sorry' (e.g. UN 1982a, 1992a; WTO 1994; Secretariat of the Convention on Biological Diversity 2000); and (b) the 'proactionary perspective' supporting the idea that 'emerging science and technology should be considered safe, economically desirable and intrinsically good unless and until it is shown to be otherwise, which means that the burden of proof is on those who want to slow down a given line of research' (More 2005).

From the definitions above, it is clear that these principles are in conflict with each other. Thus, in order to find the proper and most suitable model for dealing with risks in the field of nanotechnology, the limits of these two principles needs to be considered in greater detail.

4.4.1 The Precautionary Principle

The precautionary principle, born in the 1970s in Germany for preventing environmental damages, and subsequently broadened to include the defence of health risks, is very hard to conceptualize, due to its different interpretations.

The most common version is made up of a triple negation: 'not having evidence about a risk is not a reason for not acting preventively' (Stone 2001, p. 10799).

In trying to schematise it, Sandin affirms that it should have three main elements: (a) a hazard, (b) an uncertain threat (lack of knowledge, of full certainty about a threat), and (c) the adoption of some kind of regulation (Sandin 2006).

The first element, 'hazard', uses different qualifications of damage: it could be harmful, serious, catastrophic, irreversible and cumulative.

The second element, 'uncertain threat', refers to the lack of knowledge, i.e., the scientific uncertainty about the causality, the magnitude, the probability, and the nature of the threat. Such uncertainty is variable, but some elements of scientific

---

[5] These principles influence the way of developing the phase of 'risk assessment' and 'risk communication' as well.

knowledge about the threat and the hazard (even if this knowledge cannot be a full one) are required for grounding precautionary measures.

The anticipatory actions determine whether an intervention is required and the means to carry out the required action. These actions could be either negative (prohibitions, moratorium, etc.) or positive (intensification of investigations), but in any case they should be anticipatory.

4.4.1.1   The 'Strong Version'

'Strong' formulations of the precautionary principle can be found in the *World Charter for Nature* (UN 1982a), which affirms that: 'if potential adverse effects are not fully understood, the activities should not proceed'. It can also be found in the *Wingspread Statement* (Science and Environmental Health Network 1998), which states that: 'When an activity raises threats of harm to human health or the environment, precautionary measures should be taken even if some cause and effect relationships are not fully established scientifically […]. The proponent of an activity, rather than the public, should bear the burden of proof'.

This 'strong' version of the precautionary principle presents four main elements: the threat dimension, the uncertainty dimension, the action dimension, and the command dimension (Sandin 1999). In cases of potential and uncertain threats, an action is compulsory. So, this version imposes (a) the non usage of a new technology unless one is certain that there is no risk of harm, and (b) the adoption of a regulation whenever there is a possible harm, even if the supporting evidence is speculative and/or the economic costs of regulation are high.

This version seems to be irrational, as it embodies a general form of aversion to any kind of activity with potential risks. This makes the precautionary principle look like an anti-science, anti-technology, and anti-innovation principle (Graham 2004): indeed, looking for a 'zero risk' situation appears, in the end, completely impossible and absurd too, because 'if only actions that exposed no one to risk were permissible, the result would be a general blockade on action, which would make living together in society impossible' (Bachmann 2007, p. 9). So, this 'strong' and extreme version of the precautionary principle would bring about the prohibition and ban of any activity with inherent risks. In this sense, the precautionary principle, as applied to nanotechnology, may lead to a complete moratorium of it (ETC Group 2005; Miller 2006; NanoAction 2008) or even lead to its suspension (Joy 2000).

Furthermore, it should not be omitted that even the block or stagnation of an activity might entail some risks (Stokes 2009), or lead to the same risks that we wanted to prevent through the ban of research. Stopping the advancement of a technology does not coincide with stopping any type of risks connected with it. In fact, it is entirely possible that the 'banned' research – if let it continue – might present a solution in the prevention of the risks we wanted to avoid in the first place. This paradox is encapsulated by Sunstein (2002) and Manson (2002): if research could lead to dangerous scenarios and at the same time the absence of research could cause the same catastrophes, it means that 'the precautionary principle leads us to conclude both that we should conduct research into nanotechnology and that we should not conduct research into nanotechnology' (Clarke 2005, p. 123). In other words, the 'strong' version of the precautionary principle would be self-contradicting and lead us nowhere.

With regards to the burden of proof, which lies upon those who want to introduce a risky technology, it is apparent that such a proof of safety is a *probatio diabolica*, for none of the technologies could ever be proven safe, and in the meantime second, third, fourth order consequences may arise. The fundamental axiom of science is to prove that a negative is impossible, not just in practice, but in theory. Then, it should not be forgotten that, in reality, the claim to prove that something is safe simply means to fail to prove that it is unsafe.

4.4.1.2   The 'Anti-catastrophe' Version

In order to avoid the 'strong' version of the precautionary principle, some scholars suggest that it is only to be used in reference to catastrophic risks (Sunstein 2005; Allhoff 2009). Sunstein, for example, starts from considering that people generally fear risks and have a cognitive attitude to react against any kind of risks, and such emotional fear is the basis of the precautionary principle. Besides, there is a general misconception that nature is benevolent and human activities are always negative. So, Sunstein's proposal is to adopt a narrow 'anti-catastrophe' principle, which proposes that the precautionary principle is applied only when a particular threat creates a potentially catastrophic risk, and the existing science is not able to assign probabilities to the worst-case scenarios. Thus, it is necessary to identify all relevant risks, from both action and inaction, with the consideration of all the relevant risks related to all the considered options. However, comparing the identified risks to an unquantifiable risk of catastrophe is difficult, especially in a situation where both of the risks being compared are uncertain. This leads to the first criticism of this version, which points out the fact that in the 'knowledge' condition, there is only the mere 'possibility' of catastrophe: such concept of 'possibility' is not clear. If we mean 'possibility' as a '*logical* possibility', or if we intend it as a '*concrete and empirical*' one, in both the cases we face a conundrum: not everything that is logically possible is also empirically possible, and not everything that is concretely possible is likely to happen (Allhoff 2009).

However, one needs to consider that, if the mere hypothesis of a catastrophe is sufficient for enacting precautionary measures, then it means that this proposed 'anti-catastrophe' version of the precautionary principle requires a very low level of knowledge about the potential outcomes in order to trigger the application of the precautionary measures. Thus, the allowance of the application of the precautionary measures for any mere possibility of risk brings this interpretation of the principle near to the extreme 'strong' version, whose limitations have been outlined in the preceding section.

Another criticism is that even the ban of scientific research in view of a catastrophic risk could entail another risk again, so that 'we shall have to apply the [anti] catastrophe principle once again, negating the result of our first application' (Manson 2002, p. 273). At this point, the 'anti-catastrophe' principle would lead to the conclusion that it is better not to act at all, as each action entails risks and catastrophic effects. With this view, the principle is self-defeating, because the imposition of the remedy itself gets ruled out on grounds that it is in violation of the formulation.

4.4.1.3   The 'Weak Version'

Another possible interpretation of precautionary principle is the 'weak' one, as embedded in the Principle 15 of the Rio Declaration (UN 1992a):

> In order to protect the environment, the precautionary approach shall be widely applied by States according to their capabilities. Where there are threats [i.e., hazards] of serious or irreversible damage [i.e., harm], lack of full scientific certainty shall not be used as a reason for postponing cost-effective measures to prevent environmental degradation.

This means that is not the inaction in view of risks, but rather the choice of adopting the least risky alternative among the possible ones (according to a proportionality principle and cost-benefit analysis) before any scientific certainty of cause and effect. Despite this lack of scientific certainty, some proof of the likelihood of occurrence of harm and the severity of consequences is needed, and the burden of proof generally falls on those advocating of liability for harm (Soule 2000).

However, there are several limitations within this 'weak' version of the precautionary principle.

First of all, defining a hazard as 'serious' is vague, as this quality does not indicate any guideline about how the different risks should be ranked and how to balance between competing irreversibilities. Furthermore, it is not clear what counts as a threat of harm, as the inclusion of any potential harm or not, the means to measure harm, and the level of 'uncertainty' that is necessary to take precautionary measures are not easy to identify. Thus, it is hard to distinguish between those risks, which are deemed sufficiently probable to justify precautionary action, and those which fail to provide sufficient justification. Besides, the model of acceptable risks considers only risks and not the benefits of technology or the costs implied to reduce the risks.

4.4.1.4   The Limits of the Precautionary Principle

As seen, all the various interpretations of the precautionary principle fail to distinguish an ideal way of dealing with risks, and they do not offer any guidance at all, since they (a) do not encompass the whole quantity of risks (not considering the ones deriving from inaction), (b) fail to give attention to the benefits of a technology, (c) possess high levels of vagueness, (d) ask for impossible proofs, (e) fail to indicate the level or type of evidence of harm that is sufficient to trigger the principle, and (f) fail to indicate what level of risk is acceptable (Marchant et al. 2008).

In light of these limitations to the precautionary principle in all its incarnations, it is therefore necessary to examine another type of principle, which might be used for managing the risks of new technologies.

4.4.2   The Proactionary Principle

The proactionary principle is an ethical principle, which is elaborated as part of extropic philosophy. It was established by the Extropy Institute, a transhumanist organization that deals with meeting the technology-driven challenges and opportunities of the future (More 2006). The main vision of the institution is founded upon the need to protect the freedom of experimentation, progress and innovation, which it determines as critical to the future survival and well-being of mankind. It also adopts an attitude that allows for technology to flourish, rather than limiting its potential with an overcautious, precautionary approach. The proactionary principle is founded on the idea that, historically, all the most important technological innovations and their consequences were not so well understood at the moment of their invention, but if research had been impeded, they would never have promoted human progress.

4.4.2.1   The Limits of the Proactionary Principle

Despite its advantages, the proactionary principle also demonstrates some problematic facets. It proposes to (a) adopt an 'open door' policy to technological innovation, and (b) learn through experimentation (empiric way) and not through thinking *a priori* about something.

The proactionary principle has not got multiple strongly-articulated versions yet. Despite the fact that various aspects considered by the proactionary principle cannot be neglected and may be relevant in a management of risks, the critical aspect is that it focuses too much on the simple cause-and-effect logic, and it ignores the complex results that arise from the interactions with other developments. Moreover, the logic of learning by acting that the principle adopts would bring it up against any proposed regulation and would allow, without any limit, a complete opening to the implementation of technological development for the benefit of the society as a whole. As such, this principle perpetuates a 'radical utilitarianism [that] allows putting the individuals on the altar for the good of society' (Teshome Demissie 2008).

In addition, the reference to the cost-benefit analysis is not so feasible: since the economic quantification of the value of benefits is difficult, as the risks and benefits of new technologies are unknown and uncertain, thus making it impossible to evaluate them. In particular, it is especially difficult to determine the long-term risks, which would need a discount that is not morally acceptable (for example, discounting the value of future lives) (Kysar 2004). So, although the importance of weighing benefits and risks should not be underestimated, the centrality given to the economic analysis is sometimes problematic and does not work. It should also be considered that some benefits, such as to the environment or life or health, are valued in different ways, according to the different social and individual perceptions (Sunstein 2005).

Finally, it should be specified that, for both synbio and nanotechnology, the potential applications are heterogeneous and numerous. Thus, economic analysis should be conducted in a differentiated manner, with reference to any single application, and 'thus overwhelming available risk management resources' (Marchant et al. 2008, p. 49).

4.4.3 A Contribution from Synthetic Biology to Nanotechnology: The Notion of Responsible Stewardship

With the aforementioned limitations of the precautionary principle and proactionary principle, it is necessary at this point to find another way (Colussi 2013), which could be useful for the management of risks in the field of nanotechnology.

As a premise, it must be noted that such a position is not a hypothesis that should be distanced from the two extreme principles. Rather, this position is meant to be 'a more nuanced decision about the appropriate degree of precaution to take with respect to an emerging technology and the appropriate level and kind of support to offer it' (Murray 2011, p. 1327).

A meaningful contribution to the possible approaches in the treatment of new technologies comes from a synthetic biology report, which is elaborated by the US Presidential Commission for the Study of Bioethical Issues (PCSBI). The PCSBI is an advisory panel in the United States comprising of the nation's leaders in medicine, science, ethics, religion, law and engineering. In December 2010, the PCSBI adopted, on request of the incumbent president Obama, a report containing 18 Recommendations for a proper governance and regulation of the field. This report is based on five main ethical principles: 'the principle of public beneficence, of responsible stewardship, of intellectual freedom and responsibility, of democratic deliberation, of justice and fairness' (US PCSBI, Presidential Commission for the Study of Bioethical Issues 2010, p. 4).

With regards to biosafety, the report underlines the principles of 'public beneficence' (i.e. to act in order to maximize public benefits and minimize public harm) and of 'responsible stewardship'.[6]

---

[6] Historically, the notion of 'responsible stewardship' was almost exclusively referred to the household servant's duties for bringing food and drink to the castle's dining hall, and then it was associated with managerial skills relating to property and income; nowadays, it is applied to the commercial field (i.e., service towards passengers on ships, trains, airplanes or guests in restaurants) and, in a more recent perspective, to the environmental field, as a new type of approach to assume towards nature and biodiversity.

In terms of 'public beneficence', the PCSBI cites the *Belmont Report* (US National Commission for the Protection of Human Subjects of Biomedical and Behavioural Research 1979), a landmark statement of ethical principles for research involving human subjects, in which 'beneficence' was indicated as a core concept to require that people were treated in an ethical manner by means of respecting their decisions, protecting them from harm, and securing their wellbeing. As such, the US Commission states formally that 'we need to apply the principle of beneficence beyond the individual level, the primary emphasis of the Belmont Report, to the institutional, community, and public levels, while not overlooking possible harms and benefits to individuals. Policy makers should adopt a societal perspective when deciding whether to pursue particular benefits of synthetic biology research in the face of risks and uncertainty' (US PCSBI 2010, p. 25).

Thus, in the light of this discussion, this paper proposes the adoption of a balanced approach based on the principle of 'responsible stewardship'. This principle is connected to the approach of 'prudent vigilance', which entails an ongoing evaluation of risks along with benefits, before and after projects are undertaken. It suggests a cooperative system of information between specialised units, a preventive monitoring and control of labs, a surveillance or containment of synthetic organisms, an interaction with all stakeholders of the field, at the international and transnational level too (Anderson et al. 2012).

As for biosecurity and in the light of the so-called 'dual-use dilemma', numerous values/interests/rights are at stake, not just scientific freedom versus security. There are, indeed, more nuanced facets, such as the peril of 'anti-bioterrorism initiatives [that] will result in the unacceptable growth of government power or in unjustifiably relaxed standards for the treatment of human or animal subjects in research' (Buchanan and Powell 2010, p. 8). In fact, the measures taken by the State against bioterrorism could exacerbate the risks of biowarfare, stimulate experiments that are carried out without the consent of, or without the benefit to, the human subjects involved. It could also promote the kind of research that the government then uses for offensive purposes, resulting in more people being trained in skills that could be employed for acts of terrorism.

Thus, the importance of protecting security, and at the same time safeguarding research and controlling the behaviour of scientists and of the State itself, are relevant issues, which need to be considered.

For these reasons, the US Report stresses the importance of the responsibility of scientists, which refers again to the principle of 'responsible stewardship'. It states that synthetic biology poses some unusual potential risks, in the form of 'amateur' or 'do-it-yourself' (DIY) scientists. Therefore, 'these risks must be identified and anticipated [...] with systems and policies to assess and respond to them while supporting work toward potential benefits [...]. Responsible conduct of synthetic biology research, like all areas of biological research, rests heavily on the behaviour of individual scientists. Creating a culture of responsibility in the synthetic biology community could do more to promote responsible stewardship in synthetic biology than any other single strategy. There are actors in the world of synthetic biology [...] who practice outside of conventional biological or medical research settings. [...] This poses a new challenge regarding the need to educate and inform synthetic biologists in all communities about their responsibilities and obligations, particularly with regard to biosafety and biosecurity' (US PCSBI 2010, pp. 133–134).

The PCSBI believes that self-regulation by scientists would slow down scientific research only when the scientists perceive that there is imminent danger. This self-regulation would also prevent the establishment of a moratorium that would inappropriately limit intellectual freedom. On the contrary, 'the scientific community – in academia, government and the private sector – should continue to work together to evaluate and respond to known and potential risks of synthetic biology as this science evolves' (US PCSBI 2010, p. 145), through periodic monitoring of results by the individual and the institution.

In summary, the mentioned Report corroborates the proposal put forth in this paper that an approach that is far different from the Luddite approach towards technologies and from the *laissez-faire* approach should be adopted. This balanced approach, in the form of the principle of 'responsible stewardship', does not fight against technological and scientific advancements. Simultaneously, however, it does not allow for these advancements to proceed uncontrolled without proper regulations and guidelines to safeguard against the potential risks posed by these new technologies. Thus, congruent with this paper's proposal, the principle of 'responsible stewardship' is an approach, which adopts a balanced position, evaluates all the risks and benefits, and arranges them in a proportioned way that is required for the advancement of scientific research and the emergence of new technologies.

4.4.3.1   The Ethical and Legal Basis of Responsible Stewardship

Before the consideration of the methods with which to adopt the notion of responsible stewardship in the field nanotechnology, it is necessary to examine the ethical and legal basis of this concept.

In terms of safety and security, the notion of responsible stewardship is connected to the concept of human dignity. As defined by William Cheshire (2002): 'dignity is the exalted moral status which every being of human origin uniquely possesses. […] The possession of human dignity carries certain immutable moral obligations. These include, concerning the treatment of all other human beings, the duty to preserve life, liberty, and the security of persons, and concerning animals and nature, responsibilities of stewardship'. By this definition, 'responsible stewardship' is a 'moral obligation', the due behaviour that can 'translate' human dignity into practice. It is a behaviour of personal commitment and care that springs from the intrinsic value and inherent dignity of each human being (WYA 2002). Such a notion of dignity is intended in a collective sense, in so far as it is a feature belonging not only to the mere individual sphere, but referring to humanity as such. It includes future generations, embodying the idea that the existence and integrity of humankind as such has intrinsic worth, and therefore deserves to be protected.

More specifically, with regards to safety, responsible stewardship leans on the right to a healthy and sustainable environment, which derives from a collective notion of dignity. Indeed, the relevance of having care for each human being has the implication of having the same care for the environment in which he/she lives, since 'the protection and improvement of man's environment arises directly out of a vital need to protect human life, to assure its quality and condition, and to ensure the prerequisites indispensable to safeguarding human dignity and human worth and the development of the human personality' (Pathak 1992, p. 209). Therefore, the attention to the environment, embedded as a 'third-generation right'[7] in numerous international Declarations (UN 1972, 1992a), Conventions (UN 1982b, 1992b, c), and in some national Constitutions (e.g., South Africa), can trace its source to the notion of dignity, which emphasizes the interaction between human-kind and the environment.

In terms of security, the notion of responsible stewardship again gives application to the notion of human dignity (in order to protect people from bioterrorism or nanoterrorism). Furthermore, it is based on the protection of other human rights, such as security, public health and the freedom of research and scientific progress (which is a fundamental freedom, contained in several Constitutions all over the world: e.g., Italy and Germany).

Therefore, the principle of responsible stewardship 'reflects a shared moral obligation among members of the domestic and global communities to act in ways that demonstrate concern for those who are not in a position to represent themselves and for the environment in which future generations will flourish or suffer' (US PCSBI 2010, p. 4).

Such a notion has been also embraced by the Australian physician Bryan Furnass (Furnass 2012), who has launched the concept of 'Sustainocene' to refer to a postulated future period of over a billion years, where policy and governance structures, as well as science, technology and ethics, coordinate to achieve the social virtues of ecological sustainability and environmental integrity. In this future period, humanity will adopt an attitude of stewardship towards nature and feel morally obligated to respect biodiversity and the rights of nature itself, and to protect future inhabitants, thus changing the typically dominant anthropogenic perspective through a concerted approach summarized as the four E's, namely Enlightenment, Ecology, Education and Ethics. So, nanotechnologies and new technologies in general should be oriented by this new approach, based on stewardship, which could be applied also to environmental issues, economics, health, property, information, and so on (Chapin et al. 2009; Block 2013; Robinson et al. 2012).

---

[7]The theory of the three generations of rights was proposed by Karel Vasak (1977). This vision suggests that there is a first generation of rights represented by civil and political rights (typical of liberal societies), a second generation for economic, social and cultural life (connected with the welfare state), and a third generation that includes group and collective rights, the right to self-determination, the right to economic and social development, the right to a healthy environment, the right to natural resources, the right to communicate, the right to participation in cultural heritage, and the rights to intergenerational equity and sustainability.

4.4.3.2 'Responsible Stewardship' and Safety Risks

Since nanotechnology is part of the group of emerging technologies, the principle of 'responsible stewardship' – elaborated with reference to synbio – is perfectly applicable here. Thus, there is a need to consider its application in the field of nanotechnology.

This new approach of 'responsible stewardship' is potentially useful for scientists, as well as the public and policy makers. It allows them to weigh in on measures that could be the most suitable for the responsible development of nanotechnology.

'Responsible stewardship' could be applied to the phases of 'risk management', 'risk assessment' and 'risk communication'.

A model incorporating the notion of responsible stewardship would be aware of the fact that knowledge is a prerequisite for the exercise of responsibility. For this reason, the model starts, as in a traditional pattern, from collecting knowledge about both the risks and the benefits of the new technologies, in order to reach a comprehensive framework. At the same time, the model keeps in mind the difficulty of elaborating a full and certain knowledge about hazard characterization, exposure characterization, and beneficial aspects of applications of a new technology. As such, the model chooses to assemble a set of possible knowledge, through a constant research in risks, which is not limited to the study of possible side effects. On the contrary, it considers all reasonable alternative actions, including no action, and concentrates not only on immediate effects, but also on widely distributed and follow-on effects.

Furthermore, the research on risks is meant to be an ongoing and periodically revised process to be conducted by taking into account the interests of all potential 'victims' of risks and by involving public society and private companies. In other words, this model considers all the stakeholders in a democratic, open and transparent manner, promoting discussion and a culture of dialogue within public forums.

Thus, a dynamic, cooperative, reflexive and incremental model is proposed (Paddock 2006). It is based on the multi-stakeholder dialogue, and the broad participation aids in the understanding of the problems that arise during the 'risk assessment' phase and in the execution of possible actions during the 'risk management' phase. During the 'risk management' phase, the challenge to the model consists of finding the appropriate actions that could be proportional to the potential harms. Therefore, here the notion of 'appropriateness' is linked to the principle of proportionality, which helps to balance different interests at stake, and chooses the necessary measures in dealing with the risks with respect to human dignity, the right to the environment, and individual liberties.

As for the phase of 'risk communication', 'responsible stewardship' is able to launch (a) a joint process between stakeholders, and (b) a regular control about the state of technology and assessment of risks, thereby generating legitimacy and accountability of new technologies, and, among society, the trust in these new technologies.

4.4.3.3 'Responsible Stewardship' and Security Risks

In the field of biosecurity, the principle of 'responsible stewardship' should be adopted as the most proper approach in facing the 'dual use dilemma'. In fact, 'responsible stewardship' underlines the ethical role of (a) the scientific community and scientists towards society, (b) science publishers, (c) the 'ethical dimensions of the relationship between scientists and those in charge of national defence' (Buchanan and Powell 2010, p. 2), and (d) the responsibility of government, professional bodies and international organizations to oversee the acquisition of knowledge by scientists. This principle adopts a two-pronged approach: bottom-up and top-down (Garfinkel et al. 2007).

First of all, the principle inspires a bottom-up approach in terms of the governance of risks, by involving the scientific community in determining the soft law rules, i.e., guidelines and its own codes of conduct. Indeed, these guidelines and codes of conduct are apt for increasing the awareness of the risks posed by these new technologies and for the assignment of professionalization as a tool for governance (Weir and Selgelid 2009).

Thus, the principle of 'responsible stewardship' constitutes the basis of programs for education and training of researchers (Dando 2009), and it establishes the norms on how to manage the publications of the results of research (Journal Editors and Authors Group 2003).

The principle at stake could also guide a top-down intervention by public authorities and the State through hard law (legal and regulatory measures). The government is involved in (a) establishing the general rules for scientists (such as licenses for dealing with products or the duty to keep the State informed of developed research) and (b) in the phase of control where the sources of risk come from outside and within the State itself (in particular, by means of a decision-making authority embodying both science and security values and composed of specialists in the field).

In conclusion, the principle of 'responsible stewardship' pushes for the establishment of a culture of responsibility, and encourages the involvement of all the stakeholders in bio- and nanosecurity field, resulting in an effective scheme which encompasses the notion of 'scientific democracy' (Calvert and Martin 2009).

4.5 Concluding Remarks

Nanotechnology and synthetic biology pose numerous risks, above and beyond their possible applications. In order to deal with these risks, a solid approach is required. Instead of opting for a paralyzing principle to research, i.e., the 'strong' version of the precautionary principle, or for a *laissez-faire* approach, i.e., the proactionary approach, which could potentially open the path to abuses and distortions, a balanced solution should be adopted. With this view, the notion of 'responsible stewardship', as a principle calling for (a) an ongoing evaluation of risks, and (b) attention and care to the whole rights and interests at stake in the area of safety and security risks, is necessary. The adoption of this balanced view towards new technologies will drive the future development of subsequent emerging technologies, allowing them proceed while at the same time remaining alert in the face of their potential risks (Beyleveld and Brownsword 2012).

References

Allhoff, F. 2009. Risk, precaution, and emerging technologies. *Studies in Ethics, Law, and Technology* 3(2): 1–27.

Anderson, J., et al. 2012. Engineering and ethical perspectives in synthetic biology. *EMBO Reports* 13(7): 584–590.

Aven, T. 2008. *Risk analysis: Assessing uncertainties beyond expected values and probabilities.* Chichester: Wiley.

Bachmann, A. 2007. *Synthetic nanoparticles and the precautionary principle. An ethical analysis.* Resource document. Swiss Federal Ethics Committee on Non-Human Biotechnology. http://www.ekah.admin.ch/fileadmin/ekah-dateien/dokumentation/gutachten/e-Gutachten-Synthetische-Nanopartikel-2007.pdf. Accessed 20 Dec 2013.

Ball, P. 2005. Synthetic biology for nanotechnology. *Nanotechnology* 16(1): R01–R08.

Beck, U. 1992. *Risk society: Towards a new modernity.* New Delhi: Sage.

Beyleveld, D., and R. Brownsword. 2012. Emerging technologies, extreme uncertainty, and the principle of rational precautionary reasoning. *Law, Innovation, and Technology* 4(1): 35–65.

Bhutkar, A. 2005. Synthetic biology: Navigating the challenges ahead. *The Journal of Biolaw and Business* 8(2): 19–29.

Block, P. 2013. *Stewardship: Choosing service over self-interest.* San Francisco: Berrett-Koehler Publishers.

Boncheva, M., and G.M. Whitesides. 2004. Biomimetic approaches to the design of functional, self-assembling systems. In *Dekker encyclopedia of nanoscience and nanotechnology*, 287–294. Oxford: Taylor & Francis.

Buchanan, A., and R. Powell. 2010. *The ethics of synthetic biology: Suggestions for a comprehensive approach.* Resource document. US Presidential Commission for the Study of Bioethical Issues, PCSBI. http://bioethics.gov/cms/sites/default/files/The-Ethics-of-Synthetic-Biology-Suggestions-for-a-Comprehensive-Approach.pdf. Accessed 20 Dec 2013.

Calladine, A.M., and R.T. Meulen. 2012. Defining synthetic biology. In *Encyclopaedia of applied ethics.* Amsterdam: Elsevier.

Calvert, J., and P. Martin. 2009. The role of social scientists in synthetic biology. *EMBO Reports* 10(3): 201–204.

Cello, J., et al. 2002. Chemical synthesis of poliovirus cDNA: Generation of infectious virus in the absence of natural template. *Science* 297: 1016–1018.

Chapin III, F.S., G.P. Kofinas, and C. Folke. 2009. *Principles of ecosystem stewardship: Resilience-based natural resource management in a changing world.* Dordrecht: Springer.

Cheshire, W.P. 2002. Toward a common language of human dignity. *Ethics and Medicine* 18(2): 7–10.

Clarke, S. 2005. Future technologies, dystopic futures and the precautionary principle. *Ethics and Information Technology* 7(3): 121–126.

Colussi, I.A. 2013. Synthetic biology between challenges and risks: Suggestions for a model of governance and a regulatory framework, based on fundamental rights. *Review of Law and the Human Genome* 38: 185–216.

Dando, M. 2009. Dual-use education for life scientists? Ideas for peace and security. *Disarmament Forum* 10(2): 41–44.

De Lorenzo, V. 2008. Systems biology approaches to bioremediation. *Current Opinion in Biotechnology* 19(6): 579–589.

De Vriend, H. 2006. *Constructing life. Early social reflections on the emerging field of synthetic biology.* Resource document. Rathenau Institute. http://www.cisynbio.com/pdf/Constructing_Life_2006.pdf. Accessed 20 Dec 2013.

Deplazes, A. 2009. Piecing together a puzzle. An exposition of synthetic biology. *EMBO Reports* 10(5): 428–432.

Dietrich, J., and E. Steen. 2007. *Policy initiatives for safe realization of synthetic biology's power.* White paper. Science, Technology, and Engineering Policy Group. http://step.berkeley.edu/White_Paper/DietrichSteen.pdf. Accessed 20 Dec 2013.

Drexler, K.E. 1986. *Engines of creation.* New York: Anchor.

Eggers, J., et al. 2009. Is biofuel policy harming biodiversity in Europe? *Global Change Biology Bioenergy* 1(1): 18–34.

ETC Group. 2003. *From genomes to atoms. The big down. Atomtech: Technologies converging at the nano-scale.* http://www.etcgroup.org/content/big-down. Accessed 24 Jun 2014.

ETC Group. 2005. *A tiny primer on nano-scale technologies and 'the little bang theory'.* http://www.etcgroup.org/content/tiny-little-primer-nano-scale-technology-and-little-bang-theory. Accessed 24 Jun 2014.

European Commission. 2000. *Communication on the precautionary principle COM (2000)* 1. http://eur-lex.europa.eu/LexUriServ/LexUriServ.do?uri=COM:2000:0001:FIN:EN:PDF. Accessed 20 Dec 2013.

Feynman, R.P. 1959. There's plenty of room at the Bottom. http://www.pa.msu.edu/~yang/RFeynman_plentySpace.pdf. Accessed 20 Dec 2013.

Furnass, B. 2012. *From anthropocene to sustainocene. Challenges and opportunities.* Public lecture. Australian National University, 21 Mar 2012. http://billboard.anu.edu.au/event_view.asp?id=85103.http://www.anu.edu.au/emeritus/events/docs/From_Anthropocene_to_Sustainocene_text_only_150512.pdf. Accessed 20 Dec 2013.

Garfinkel, M.S., et al. 2007. *Synthetic genomics: Options for governance.* J. C. Venter Institute. http://www.synbiosafe.eu/uploads///pdf/Synthetic%20Genomics%20Options%20for%20Governance.pdf. Accessed 20 Dec 2013.

Gervais, D. 2010. The regulation of inchoate technologies. *Houston Law Review* 47(3): 666–705.

Gibson, D.G., J.I. Glass, C. Lartigue, V.N. Noskov, R.-Y. Chuang, et al. 2010. Creation of a bacterial cell controlled by a chemically synthesized genome. *Science* 329: 52–56.

Glass, J.I., N. Assad-Garcia, N. Alperovich, S. Yooseph, M.R. Lewis, M. Maruf, et al. 2006. Essential genes of a minimal bacterium. *Proceedings of the National Academy of Sciences of the United States of America* 103(2): 425–430.

Gouvernement du Québec. 2006. *Position statement. Ethics and nanotechnology: A basis for action.* Summary, recommendations and commentaries. Commission de l'éthique de la science et de la technologie. http://www.ethique.gouv.qc.ca/fr/assets/documents/Nano/AvisNano_EN.pdf. Accessed 24 Jun 2014.

Graham, J.D. 2004. *The perils of the precautionary principle: Lessons from the American and European experience.* The Heritage Foundation. http://www.heritage.org/research/lecture/the-perils-of-the-precautionary-principle-lessons-from-the-american-and-european-experience. Accessed 20 Dec 2013.

Grunwald, A. 2005. Nanotechnology – A new field of ethical inquiry? *Science and Engineering Ethics* 11: 187–201.

High-Level Expert Group of the European Commission. 2005. *Synbiology. An analysis of synthetic biology research in Europe and North America.* http://www2.spi.pt/synbiology/documents/SYNBIOLOGY_Literature_And_Statistical_Review.pdf. Accessed 20 Dec 2013.

Hobom, B. 1980. Gene surgery: On the threshold of synthetic biology. *Medizinische Klinik* 75(24): 834–841.

Journal Editors and Authors Group. 2003. Joint statement on scientific publication and security. Uncensored exchange of scientific results. *Proceedings of the National Academy of Sciences of the United States of America* 100(4): 1464.

Joy, B. 2000. Why the future doesn't need us. *Wired Magazine.* http://www.wired.com/wired/archive/8.04/joy_pr.html. Accessed 20 Dec 2013.

Kirby, J.R. 2010. Synthetic biology: Designer bacteria degrades toxin. *Nature Chemical Biology* 6(6): 398–399.

Kysar, D.A. 2004. Climate change, cultural transformation, and comprehensive rationality. *Boston College Environmental Affairs Law Review* 31(3): 555–590.

Manson, N.A. 2002. Formulating the precautionary principle. *Environmental Ethics* 24(3): 263–274.

Marchant, G.E., D.G. Sylvester, and K.W. Abbott. 2008. Risk management principles for nanotechnology. *NanoEthics* 2(1): 43–60.

Mast, E.E., and J.W. Ward. 2008. Hepatitis B vaccines. In *Vaccines*, ed. S. Plotkin, W. Orenstein, and P. Offit, 205–242. Philadelphia: Saunders.

Maynard, A.D. 2006. Nanotechnology: The next big thing, or much ado about nothing? *Annals of Occupational Hygiene* 51(1): 1–12.

Miller, G. 2006. *Nanomaterials, sunscreens, and cosmetics: Small ingredients, big risks.* Resource document. Friends of the Earth. http://nano.foe.org.au/sites/default/files/FoEA%20nano%20cosmetics%20report%202MB.pdf. Accessed 20 Dec 2013.

More, M. 2005. *The proactionary principle.* http://www.ethique.gouv.qc.ca/fr/assets/documents/Nano/AvisNano_EN.pdf. Accessed 24 june 2014.

More, M. 2006. Proactionary nano-policy: Managing massive decisions for tiny technologies. *The Journal of Geoethical Nanotechnology.* http://www.terasemjournals.org/GNJournal/GN0102/more_01a.html. Accessed 20 Dec 2013.

Murray, T.H. 2011. What synthetic genomes mean for our future: Technology, ethics, and law, interests and identities. *Valparaiso University Law Review* 45(1): 1315–1342.

NanoAction. 2008. *Principles for the oversight of nanotechnologies and nanomaterials.* http://nanoaction.org/nanoaction/doc/nano-02-18-08.pdf. Accessed 20 Dec 2013.

O'Malley, M.A., A. Powell, J.F. Davies, and J. Calvert. 2008. Knowledge-making distinctions in synthetic biology. *Bioessays* 30(1): 57–65.

Organization for Economic Cooperation and Development (OECD). 2003. *Emerging risks in the 21st century: An agenda for action.* http://www.oecd.org/dataoecd/20/23/37944611.pdf. Accessed 20 Dec 2013.

Paddock, L. 2006. Keeping pace with nanotechnology: A proposal for a new approach to environmental accountability. *Environmental Law Reporter News and analysis* 36(12): 10943–10952.

Pathak, R.S. 1992. The human rights system as a conceptual framework for environmental law. In *Environmental change and international law*, ed. E.B. Weiss, 205–243. Tokyo: United Nations University Press.

Rawls, R. 2000. 'Synthetic biology' makes its debut. *Chemical and Engineering News* 78(17): 49–53.

Ro, D.K., et al. 2006. Production of the antimalarial drug precursor artemisinic acid in engineered yeast. *Nature* 440: 940–943.

Robinson, J.S., M.S. Walid, and A.C.M. Barth. 2012. *Toward healthcare resource stewardship: Health care issues, costs, and access.* New York: Nova Science.

Roco, M. 2006. *Risk governance for nanotechnology.* IRGC workshop. http://www.irgc.org/IMG/pdf/Mike_Roco_Risk_Governance_for_Nanotechnology_.pdf. Accessed 20 Dec 2013.

Sandin, P. 1999. Dimensions of the precautionary principle. *Human and Ecological Risk Assessment* 5(5): 889–907.

Sandin, P. 2006. A paradox out of context: Harris and Holm on the precautionary principle. *Cambridge Quarterly of Healthcare Ethics* 15(2): 175–183.

Savage, D.F., J. Way, and P.A. Silver. 2008. Defossiling fuel: How synthetic biology can transform biofuel production. *American Chemical Society Chemical Biology* 3(1): 13–16.

Schmidt, M. 2009. Do I understand what I can create? Biosafety issues in synthetic biology. In *Synthetic biology. The technoscience and its societal consequences*, ed. M. Schmidt, A. Kelle, A. Ganguli, and H. De Vriend, 81–100. New York: Springer.

Science and Environmental Health Network. 1998. *Wingspread consensus statement on the precautionary principle.* http://www.sehn.org/wing.html. Accessed 20 Dec 2013.

Scrinis, G., and K. Lyons. 2010. Nanotechnology and the techno-corporate agri-food paradigm. In *Food security, nutrition and sustainability: New challenges, future options*, ed. G. Lawrence, K. Lyons, and T. Wallington, 252–270. London: Earthscan.

Secretariat of the Convention on Biological Diversity. 2000. *Cartagena protocol on biosafety to the convention on biological diversity: Text and annexes.* http://bch.cbd.int/protocol/text/. Accessed 20 Dec 2013.

Serrano, L. 2007. Synthetic biology: Promises and challenges. *Molecular Systems Biology* 3: 158.

Slovic, P. 2000. *The perception of risk.* London: Earthscan.

Soule, E. 2000. Assessing the precautionary principle. *Public Affairs Quarterly* 14: 309–328.

Stokes, E. 2009. Regulating nanotechnologies: Sizing up the options. *Legal Studies* 29(2): 281–304.

Stone, C.D. 2001. Is there a precautionary principle? *Environmental Law Repertory* 31(7): 10790.

Sunstein, C.R. 2002. *Risk and reason: Safety, law and the environment.* Cambridge: Cambridge University Press.

Sunstein, C.R. 2005. *Laws of fear: Beyond the precautionary principle.* Cambridge: Cambridge University Press.

Swiss Re. 2004. *Nanotechnology: Small matter, many unknowns.* http://www.temas.ch/IMPART/IMPARTProj.nsf/35370CF58146AD0FC125736300496D94/$FILE/SwissRe_Nano_en.pdf?OpenElement&enetarea=02. Accessed 24 Jun 2014.

Synthetic Genomics Inc. 2009. *Agricultural products.* http://www.syntheticgenomics.com/what/agriculture.html. Accessed 20 Dec 2013.

Szostak, J.W., D.P. Bartel, and P.L. Luisi. 2001. Synthesizing life. *Nature* 409: 387–390.

Szybalski, W. 1974. In vivo and in vitro initiation of transcription. *Advances in Experimental Medicine and Biology* 44(1): 23–24.

Teshome Demissie, H. 2008. Taming matter for the welfare of humanity: Regulating nanotechnology. In *Regulating technologies. Legal futures, regulatory frames and technological fixes*, ed. R. Brownsword and K. Yeung, 327–356. Oxford/Portland: Hart Publishing.

The Royal Society. 1992. *Risk, analysis, perception, management.* London: The Royal Society.

The Royal Society, and The Royal Academy of Engineering. 2004. *Nanoscience and nanotechnologies: Opportunities and uncertainties.* London: The Royal Society.

Tumpey, T.M., et al. 2005. Characterization of the reconstructed 1918 Spanish influenza pandemic virus. *Science* 310: 77–80.

UK Parliamentary Office of Science and Technology. 2009. *The dual-use dilemma.* http://www.parliament.uk/documents/post/postpn340.pdf. Accessed 20 Dec 2013.

United Nations. 1972. *Stockholm declaration.* Conference on the human environment. http://hqweb.unep.org/. Accessed 30 Mar 2012.

United Nations. 1982a. *World charter on nature* (G.A. Res. 37/7. 11, UN Doc). A/RES/37/7. http://www.un.org/documents/ga/res/37/a37r007.htm. Accessed 20 Dec 2013.

United Nations. 1982b. *Conventions on the law of the sea.* http://www.un.org/depts/los/convention_agreements/convention_overview_convention.htm. Accessed 20 Dec 2013.

United Nations. 1992a. *Rio declaration on environment and development.* Conference on environment and development. http://www.unep.org/Documents.Multilingual/Default.asp?documentid=78&articleid=1163. Accessed 20 Dec 2013.

United Nations. 1992b. *Framework convention on climate change.* Conference on environment and development. http://unfccc.int/2860.php. Accessed 20 Dec 2013.

United Nations. 1992c. *Convention on biological diversity.* Conference on environment and development. http://www.cbd.int/. Accessed 20 Dec 2013.

US National Commission for the Protection of Human Subjects of Biomedical and Behavioural Research. 1979. *The Belmont report: Ethical principles and guidelines for the protection of human subjects of research.* http://www.hhs.gov/ohrp/humansubjects/guidance/belmont.html. Accessed 24 Jun 2014.

US National Science and Technology Council. 1999. *Nanotechnology research directions.* IWGN workshop report. http://www.wtec.org/loyola/nano/IWGN.Research.Directions/IWGN_rd.pdf. Accessed 20 Dec 2013.

US Presidential Commission for the Study of Bioethical Issues, PCSBI. 2010. *The ethics of synthetic biology and emerging technologies*. Washington, DC: US Presidential Commission.

Vasak, K. 1977. *Human rights: a thirty-year struggle: The sustained efforts to give force of law to the universal declaration of human rights*. Paris: UNESCO.

Weir, L., and M.J. Selgelid. 2009. Professionalization as a governance strategy for synthetic biology. *Systems and Synthetic Biology* 3(1–4): 91–97.

Whittall, H. 2009. The ethics of synthetic biology. In *Ethical aspects of synthetic biology*, ed. The European Group on Ethics in Science and New Technologies to the European Commission, 27. Luxembourg: Publications Office of the European Union.

World Health Organization (WHO). 2006. *Biorisk management: Laboratory biosecurity guidance*. Geneva: World Health Organization.

World Trade Organization (WTO). 1994. *Agreement on the application of sanitary and phytosanitary measures (SPS Agreement)*. http://www.wto.org/english/tratop_e/sps_e/spsagr_e.htm. Accessed 20 Dec 2013.

World Youth Alliance (WYA). 2002. *Declaration on responsible stewardship for the world summit on sustainable development*. http://www.wya.net/getinvolved/declarationsandstatements/declarationonresponsiblestewardship.html. Accessed 20 Dec 2013.

Part II
Public Engagement and Technology Assessment

Chapter 5
Technology Assessment Beyond Toxicology – The Case of Nanomaterials

Torsten Fleischer, Jutta Jahnel, and Stefanie B. Seitz

5.1 Introduction

Although the term technology assessment (TA) is generic for non-uniform, partly even contradictory approaches and activities, it can be defined as 'a *scientific, interactive and communicative process which aims to contribute to the formation of public and political opinion on societal aspects of science and technology*' (Decker and Ladikas 2004). Regarding responsibility in nanotechnology development TA activities aim to provide knowledge which politics and society can use as a basis for action and decision-making in the governance process of nanotechnology. Within this process the focus was put on manufactured particulate nanomaterials (MPNs), a group of substances – inseparably linked to nanotechnology – that are only commonly characterized by their nano-size. The peculiarity of MPNs is that their properties differ significantly from those of lager particles of the same material. This makes them suitable for new or improved applications which are expected to be a major opportunity for the economic and sustainable development of many countries. However, these new properties deriving from the nano-size are just the same as those which concern scientists, in particular, but also policy makers, a number of stakeholders and parts of the general public.

Experiences of the past, e.g. with chemicals, asbestos or ultrafine particles, showed that new materials may be a source of new threats for human health and the environment (Oberdörster et al. 2005). The scientific community – in particular the toxicologists – was and is still expected to answer the question of whether MPNs pose environmental, health and safety (EHS) risks or not, and to provide policy makers with the appropriate knowledge to perform risk assessment as a prerequisite for science-based risk management. Beside the problem that the (nano)toxicology

T. Fleischer (✉) • J. Jahnel • S.B. Seitz
Institute for Technology Assessment and Systems Analysis (ITAS), Karlsruhe Institute
of Technology (KIT), Karlsruhe, Germany
e-mail: torsten.fleischer@kit.edu

S. Arnaldi et al. (eds.), *Responsibility in Nanotechnology Development*, The International
Library of Ethics, Law and Technology 13, DOI 10.1007/978-94-017-9103-8_5,
© Springer Science+Business Media Dordrecht 2014

research agenda is not only driven by the aim to produce systematically knowledge for decision-making of politics and society, the current concept for assessing nano-specific risks is the conventional expert-based chemical risk assessment procedure (SCENIHR 2007, 2009). This concept is limited to a narrow toxicological perspective defining risk itself as a hazard multiplied by exposure. The toxicological risk assessment paradigm is based on confidence in the knowledge used despite serious methodological uncertainties in the case of nanotechnology. Accordingly, a wider concept is needed which allows for a plurality in perspective, actors and different kinds of knowledge adequately considering societal impacts for understanding risk in a broader sense than simply experts. In addition, regulation based on quantitative risk assessment is an inherently slow governance process. This leads to alternative more adaptive governance frameworks, such as those suggested by the Environmental Defense Fund and Dupont (Nano Risk Framework, Environmental Defense Fund and Dupont 2007) or the International Risk Governance Council (IRGC 2006). The IRGC framework combines the scientific risk-benefit assessment with an assessment of risk perception and the societal context of risk, called concern assessment. Here we deal with the outcomes of the different methods for concern assessment. We will discuss the possible support for an inclusive understanding of risk appraisal as a precondition for responsible risk management and risk governance.

5.2   Why Beyond Toxicology?

5.2.1   Limitations of the Classical Risk Assessment

Toxicology as discipline aims to study the adverse effects of chemicals on living organisms, especially humans. Thus, it also provides knowledge for decision-making during risk management. Moreover, toxicologists were among the first who expressed concerns regarding potential risks of MPNs towards health and the environment. Subsequent, nanotoxicology emerged from the classical toxicology, and studies in particular the biological effects of engineered nanomaterials on living organisms and in ecosystems (Oberdörster 2010a). In general, (nano)toxicology is the justification for risk governance according the precautionary principle. Thus, nanotoxicology research is incorporated into risk assessment as a part of the governance of MPN-risks.

The classical risk assessment is a well-established and formalized process intended to

> calculate or estimate the risk to a given target organism, system or (sub)population, including the identification of attendant uncertainties, following exposure to a particular agent, taking into account the inherent characteristics of the agent of concern as well as the characteristics of the specific target system (OECD 2003).

The risk assessment process consists of four steps: hazard identification, hazard characterization (usually summarized as hazard assessment), exposure assessment, and risk characterization.

According to the Risk Commission (2003), a scientific risk assessment process primarily deals with consequences of the effects of noxious agents to human health. Risk assessment resembles a process in which the probability of a harmful effect on individuals or populations is quantified. The framework was developed for conventional chemicals as an information and decision-supporting tool for possible regulations. Associated uncertainty in the progress is managed by the application of safety factors. There is a consensus that the classical measures of toxicology are in principle applicable to nanomaterials, but standard procedures of risk assessment have to be modified (e.g. Rocks et al. 2008).

The EU Scientific Committee on Emerging and Newly Identified Health Risks (SCENIHR) stated already in their 2007 opinion that the current methodologies are generally likely to be able to identify the hazards associated with the use of nanomaterials. However, they see the need for modifications for the guidance on the assessment of risks (SCENIHR 2007, 2009). Moreover, assessing risks of nanomaterials using conventional paradigms may not be sufficient to capture all the dimensions of risk of an active nano-bio material, as risk may arise not only from its inherent material toxicity but also from its interactions with complex biological systems (Maynard et al. 2006).

The main limitations for current procedures to assess the risks of nanomaterials are:

- The question of the identification and definition of the term 'nanomaterial' poses a challenge for framing a 'substance class' with a high diversity for risk assessment.
- Equipment and methods for characterization and detection of nanomaterials are often not appropriate and need further optimization (Maynard et al. 2006; Tiede et al. 2008; Marquis et al. 2009; Leach et al. 2011). It is still impossible to detect nanomaterials in biological matrixes.
- A definition or concept for dose/concentration is still missing.
- High quality exposure and dosimetry data is also still missing. Many exposure-related studies are published on occupational scenarios while many fewer studies are published on environmental and consumer exposure as well as about both acute and chronic exposures (ENRHES 2010; Aschberger et al. 2011).
- Standardized methods including appropriate controls are largely still missing. Exacerbating factors – such as surface functionalization, dispersing behaviour in biological media or the use of solvents in the case of non-dispersing nanoparticles (e.g. fullerenes) in aqueous media – that may e.g. produce testing artefacts (Henry et al. 2007) – are not addressed sufficiently in many studies (ENRHES 2010; Aschberger et al. 2011).
- Studies that showed no significant (hazardous) effects are usually not published, even though they are crucial to relieve MPNs from the suspicion of hazard (Krug and Wick 2011).

- There is an ongoing debate on the significance of high-dose *in vitro* or *in vivo* studies conducted so far and whether or not the used methods are suitable for hazard characterization (e.g. Oberdörster 2010b).
- For eco-toxicological studies it is, in general, difficult to simulate real environmental scenarios since the dose is quite unknown and the extrapolation of data is very limited (ENRHES 2010; Aschberger et al. 2011).

Limitations within the classical risk assessment processes concerning uncertainties and knowledge gaps that also occur with other chemical substances become even more overt in the case of MPNs. Moreover, one overarching difficulty will most probable always remain: In contrast to the vast majority of substance classes of hazardous chemicals that need to undergo risk assessment, MPNs share no common characteristics apart from the fact that the primary particles are in nano-scale. Although there are a number of approaches to categorize MPNs in a kind of 'hazard classes' or develop EHS risk prediction systems (e.g. Foss Hansen et al. 2007; Xia et al. 2009, 2010; Burello and Worth 2011; Puzyn et al. 2011), it is the consensus in the nanotoxicology community that due to the knowledge gaps and intrinsic limitations of characterization of MPNs, today only a 'case-by-case' assessment is responsible and sound. Thus, risk assessment of MPNs requires the full dataset for each and every kind of MPN. This makes the progress of gathering the relevant data for this case-by-case approach extremely slow – although the literature body is increasing constantly. Therefore also today, a complete risk assessment is only possible for a small selection of highly abundant MPNs, like nano-silver, carbon nanotubes and fullerenes, or titanium dioxide nanoparticles (e.g. Krug and Wick 2011; Aschberger et al. 2011).

Additionally classical risk assessment favours scientific knowledge that can be measured, weighted and monitored. This ignores the importance of values, ethics and tacit forms of knowledge when judging risks. There are several inherent value judgements in risk assessment and value-based decisions should be left to the political decision-makers (Senjen and Hansen 2011).

5.2.2   On the IRGC Risk Management Framework

Facing these limitations of the risk assessment there was the request for a more holistic approach beyond the expert-based chemical risk assessment procedure. In order to consider societal impacts and societal needs for understanding risk in a broader sense than experts, the classical toxicological-driven risk assessment paradigm should be widened.

In its white paper published in 2006, the International Risk Governance Council tackled this problem and introduced a new conceptual framework for the risk governance of nanotechnology (IRGC 2006). Risk Governance, according to the IRGC includes the totality of actors, rules, conventions, processes, and mechanisms concerned with how relevant risk information is collected, analysed and communicated and management decisions are taken. Encompassing the combined risk-relevant decisions and actions of both governmental and private actors, risk governance is of particular importance in, but not restricted to, situations where there is no single authority to take a binding risk management decision but where instead the nature of the risk requires the collaboration and co-ordination between a range of different stakeholders. Risk governance, however, not only includes a multifaceted, multi-actor risk process but also calls for the consideration of contextual factors such as institutional arrangements (e.g. the regulatory and legal framework that determines the relationship, roles and responsibilities of the actors and co-ordination mechanisms such as markets, incentives or self-imposed norms) and political culture including different perceptions of risk (Renn 2008).

Concerning the responsibility in nanotechnology development, the IRGC framework is a sophisticated risk management model. It involves a multitude of different actors in a dynamic process with various iterations and feedbacks. It acknowledges that risk governance decisions have to be taken in instances of complexity, uncertainty and ambiguity. Therefore, strategies should be based on a corrective and adaptive approach and take into account the level and extent of available knowledge and a societal balancing of the predicted risks and benefits. The framework includes two innovative concepts for the governance of (potential) risks arising from the use of MPNs:

- It integrates a scientific risk-benefit assessment (including environment, health, and safety (EHS) and ethical, legal, and other social issues (ELSI)), with an assessment of risk perception and the societal context of risk (referred to in the white paper as concern assessment).
- Inherent is the need for all interested parties to be effectively engaged, for risk to be suitably and efficiently communicated by and to the different actors and for decision-makers to be open to public concerns.

The IRGC Framework is a cyclical process and consists of four phases: The 'Pre-Assessment' which can be seen as the trigger or initiator of the whole assessment and management process is the first phase. Subsequent, 'Risk Appraisal' as the second phase of the IRGC risk governance framework follows and comprises two elements: risk assessment (see Sect. 5.2.1.) and concern assessment (see Sect. 5.2.3.). This is followed by the third phase called 'Tolerability and Acceptability Judgment' which brings together the classic risk characterization and risk evaluation as a new element. Finally, 'Risk Management' (Phase 4) has to react not only to new scientific results regarding a hazard or an exposure to it. It also reacts to changing societal or cultural factors like altering expectations on risk reduction procedures, new judgments about tolerability and acceptability of risks, developing value systems or shifting risk perceptions of different actors.

Some authors criticised this framework because public participation is still perceived as a factual input, as part of an expert-driven process, rather than empowerment of citizens (Senjen and Hansen 2011).

5.2.3 The Role of Concern Assessment

During the risk management phase, one has to address what the concerns of the general public and the stakeholders are, when it comes to a widespread market introduction and usage of MPNs. In short: within a risk governance process that considers the political and institutional conditions in modern societies, risk assessment has to be complemented by a concern assessment.

In a book article that addresses conceptual issues of the IRGC framework raised by external experts in a round of formal comments, the lead authors define concern assessment as

> a social science activity aimed at providing sound insights and a comprehensive diagnosis of concerns, expectations and perceptions that individuals, groups or different cultures may link to the hazard (Renn and Walker 2008).

Understanding these different concerns, expectations and perceptions is an important factor in getting to know better how individuals and groups perceive and assess risks and what actions (or non-actions) are perceived as being risky for what reasons. In addition, it helps to comprehend how the different actors are expected to develop and implement adequate measures in risk management and risk communication. Investigations of the evolving socio-cultural and political context in which research at the nano-scale is conducted, the societal needs that nanotechnology may satisfy and the popular images that experts, politicians and representatives of the various publics associate with nano-science and nanotechnology are additional elements in improving the societal knowledge about adequate risk management procedures (IRGC 2006).

Fundamental for the comprehensive diagnosis of concerns is the meaning of risk. According to IRGC (2005) and Renn and Walker (2008), risk is characterized in general as a '*mental construction*', which means that risk is

> not a real phenomena but originates in the human mind. Actors, however, creatively arrange and reassemble signals that they get from the 'real world' providing structure and guidance to an ongoing process of reality enactment. So risks represent what people observe in reality and what they experience.

Generally speaking, the perception of technological risks depends on two sets of factors. The first consists of psychological factors such as perceived threat, familiarity, personal control options and positive risk-benefit ratio. The second set includes political and cultural factors such as perceived equity and justice, visions about future developments and effects on personal interests and values. While the first set of components can be predicted to some degree on the basis of the properties of the technology itself and the situation of its introduction, the second set is almost impossible to predict (IRGC 2006).

While conventional chemical risk assessment can build upon a long tradition of scientific discussion, methodological development and established organizational and institutional practices, concern assessment is still in its early stages. That notwithstanding, what is needed is a systematic assessment of the concerns and preferences of the various actor groups and the public at large, together with a systematic feedback of its results to the related regulatory and legislative processes. These are necessary prerequisites to improve our understanding of the likely societal responses to the developments in nanomaterials and nanotechnology. This is also important for the implementation of risk governance structures that are accepted as socially responsible and to avoid public controversies and potential conflicts.

5.3 How to Translate Concern Assessment into Praxis?

5.3.1 Methodological Challenges

In the IRGC framework, risk communication has a central role and several functions. First of all, this should enable an information flow between the different players (policy makers, scientists of the different disciplines, stakeholders and representatives of the general public) as well as the different phases of the process. Moreover, risk communication is the key to building trust for the risk management process and improving the performance of the management system significantly (IRGC 2006). Concerning the communication between persons that are professionally involved in the process (scientists, policy makers) and 'the outside world', another principal function of risk communication is to enable concerned citizens to make their own balanced risk-judgment. This means that any person or social group affected by risks should be sufficiently well-informed to make a personal judgment of the risks, which meets their own criteria.

Thus, the aim of dialogues, engagement and participation events engaged in concern assessment and/or risk communication should be to address fundamental issues and characteristics of the risk problem, such as the degree of complexity, the nature of uncertainty and ambiguity. High levels of ambiguity require the most inclusive strategy for participation since not only directly affected groups but also those indirectly affected have something to contribute to a debate. To translate these rather abstract requirements into actual political action remains a demanding task.

One of the key problems in developing formats for public participation is that the general public – by definition – is neither organized, nor can it be represented adequately by self-appointed representatives. To address this problem, a number of innovative tools such as consensus conferences, citizens' juries, focus groups, scenario workshops, etc. which are more dialogue-oriented than the classic forms (like exhibiting documents for inspection and providing opportunities to submit comments) and made for more effective participation by non-organized citizens have been developed and tested, and numerous experiences regarding the design of participatory procedures have been acquired (e.g. Gavelin et al. 2007; Hullmann 2008; Bonazzi 2010).

Among a set of well-established methods that social science used and uses to study perceptions of nanotechnology's benefits and risks within individuals, groups or the society as a whole are quantitative and qualitative methods. Each of them has its own pros and cons. Quantitative methods – including surveys which are designed to ascertain large and therefore representative datasets as well as experimental studies using non-probability samples – for example, allow for testing and revising existing hypotheses, and making statements about defined groups of people. Typical examples are large, standardized polls within a representative sample of a population. In contrast, qualitative methods are rather designed to gain insights into individual arguments, ideas or values and to explore new aspects of an issue. Thus, they are designed rather open (not standardized) to capture even unexpected facts. Beside in-depth interviews, focus groups are typical examples of qualitative methods (Fleischer and Quendt 2007; Fleischer et al. 2012a).

Generally speaking, the landscape of research into perceptions of nanotechnology and nanomaterials – and the related concerns – among European citizens is somewhat patchy. To our knowledge, representative studies about the familiarity with, attitudes towards and perceptions of nanotechnology covering all member states have only been performed within three Special Eurobarometer surveys in 2002, 2005 and 2010. This research has been complemented with a number of country studies over the last few years (e.g. BMRB 2004; BfR 2008). Since these surveys have used various methodologies and mostly different questions or different question wordings, their results are hard to compare with each other and with the Eurobarometer findings. In the following chapters the quantitative and qualitative results are discussed separately.

5.3.2   Quantitative Results: Eurobarometer Survey 2010

The most recent – and most reliable – representative data on the awareness, expectations and attitudes of the general public towards nanotechnology in Europe can be taken from a 2010 Special Eurobarometer survey on biotechnology (Eurobarometer 2010). This survey covers a representative sample of the population of the respective nationalities of the European Union Member States (plus Iceland, Norway, Switzerland, Croatia and Turkey), resident in each of the Member States and aged 15 years and over. The survey was carried out between the 29th of January and the 17th of February 2010. All respondents were interviewed face-to-face in people's homes and in the appropriate national language. The sample size (usually around 1,000 respondents per country) permits accuracy (confidence interval) of ca. ±3 percent points. The Eurobarometer study allows for comparing public opinions in different EU Member and Associated States. It gives some indications of what effects public dialogue and engagement exercises may have had on public opinion in a particular country prior to the opinion poll (ObservatoryNano 2012).

Regarding nanotechnology, respondents had been asked first if they had ever heard of nanotechnology before. Forty-six percent of Europeans had heard of nanotechnology, while 54 % had never heard of it. Looking at the socio-demographic data, they show that gender, education and age are factors. Fifty-four percent of men (compared to 39 % of women) had heard of nanotechnology. Most likely to have heard of nanotechnology were managers (76 %), students (60 %) or self-employed people (57 %) as well as persons who left full-time education age 20+ (68 %) and everyday users of the internet (62 %). Least familiar with nanotechnology were house persons (30 %), retired (35 %) or unemployed (38 %) people as well as those who left school at age 15 or below (22 %) and non-users of the internet (25 %). Forty-one percent of Europeans expected a positive impact of nanotechnology on their way of life in the next 20 years, 40 % did not know, 10 % expected a negative effect and 9 % thought that nanotechnology would have no effect.

In order to tap into perceptions of, expectations of and concerns about nanotechnology, respondents were presented ten statements about nanotechnology and asked whether they totally agreed, tended to agree, tended to disagree or totally disagreed. The statements covered four clusters: perceived benefit, perceived safety/risk, perceived fairness/unfairness with regard to distributional equity and worries related to unnaturalness.

As a general impression at the European level, one third of the respondents believed that nanotechnology may do harm to the environment, is not safe to human health and is not safe to future generations, respectively. One third expressed an opposite view and one third did not know. A more regional perspective showed interesting differences: The higher the number of respondents in a certain region that had already heard about nanotechnology, the higher the number of respondents that didn't agree that nanotechnology is safe to their health and agreed that nanotechnology would do harm to the environment (Fleischer et al. 2012a). On this highly aggregated level, there seems to be a positive correlation between perceived knowledge about and perceived risk of nanotechnology, an observation that has to be confirmed by future in-depth research.

Surprisingly, in a number of countries, the percentage of respondents who express an opinion about perceived safety/risk of nanotechnology is even higher (statistically significant) than the percentage of respondents that have already heard about nanotechnology. In other words, the perceptions of some respondents appear to be based on factors other than factual knowledge about nanotechnology.

A more detailed analysis was provided by Gaskell et al. in an accompanying report to the Eurobarometer survey, presenting research from the FP7 project 'Sensitive Technologies and European Public Ethics' (STEPE). They found that, across the European public

> the balance of opinion is that nanotechnology is somewhat more likely to be beneficial than not, to be unsafe rather than safe, to be inequitable rather than equitable, and not particularly worrying (though, equally, not particularly unworrying) (Gaskell et al. 2010).

They also showed that perceived safety is by far the most influential variable on overall support of or opposition to nanotechnology, followed by benefit, worries related to unnaturalness and lastly inequity.

5.3.3 Qualitative Results: Observations in Public Engagement Exercises and in Dedicated Focus Group Studies

Additional insights for studying perceptions and concerns related to nanoparticles can be gained from the results of qualitative methods. Various participatory projects – like *NANOBIO-RAISE, DEEPEN, TIME for Nano*, German *NanoCare*, Austrian *Risiko:dialog*, Danish Survey of 2004, UK *Nanotechnology, Risk and Sustainability*, Dutch *Nanopodium*, or Swiss *Publifocus*) have included qualitative methods such as interviews or focus groups.

Only one of the projects that was using qualitative methods was focused explicitly on conceptions and concerns regarding MPNs: the focus groups that were conducted within the German 'NanoCare' project (Fleischer and Quendt 2007), while the remainder were dealing with nanotechnology in general. Two further focus group discussions dealing with MPNs were performed within the NanoSafety project (Fleischer et al. 2012a). Both events focused on getting insights into ideas, concepts and associations that citizens had of nanoparticles and nanoproducts. A second part of the discussions addressed the participants' expectations regarding political action.

During focus group discussions the citizens talked with each other using statements, narratives, comparisons, analogies, metaphors and stories. With these verbal tools they expressed only indirectly accessible mental and cognitive constructs like concerns, perceptions, opinions and expectations. Furthermore, underlying reasons, rationalities but also individual emotional reactions and feelings were expressed.

The vast majority of people still have little or no idea of what nanotechnology is or about its possible implications. Despite this, members of the public have already expressed similar concerns to those associated with other technologies perceived as being risky, particularly around governance structures and corporate transparency. Many citizens were astonished about the broad scope, spectrum and extent of 'nano-products' already available. Many discussants arbitrarily mixed their terminology and used nanoparticles, nanotechnology and sometimes also 'nanoproducts' quasi synonymously.

The different concerns and expectations of the participants were motivated by special individual contexts and could be linked to concrete needs and intentions. The statements during the events of different qualitative methods were grouped according to the following main dimensions.

Regarding *human health*, improvements in disease prevention, early disease detection or medical treatment were expected. The participants hoped to benefit from improved medicinal applications (nanomedicine). Thereby, they were concerned about potentially adverse health effects (mainly due to inhalation) of MPNs, the entry of MPNs into the human body due to their very small size and scientific uncertainty regarding the behaviour of nanoparticles in the human body as well as uncertainties with regard to risk assessment.

Improvements due to MPN applications were expected for the *environment*, such as effects on pollution prevention and remediation, also energy conservation, efficiency gains in production due to miniaturization effects, cleaner manufacture with fewer emissions and less waste. Nanotechnology-based environmental technology applications like devices for waste water treatment were expected to bring benefits as well as the substitution of classical hazardous chemicals. However, the participants were concerned about uncontrolled release of MPNs into the environment, their possible occurrence in ground water, in air and their possible enrichment in the food chain. They worried about life-cycle impacts like energy and resource intensive manufacturing, problems in the recycling and disposal phases, especially considering disposal and behaviour in wastewater treatment.

Asked for their *acceptance* of MPNs, the citizens were less reluctant to the use in medical applications, cosmetics and other sectors. They appreciated consumer and household 'nanoproducts' increasing convenience in daily lives. The contribution to the progress of medical applications and the possible substitution of chemicals of concern was stated as an advantage, too. Nevertheless, they argued that due to the lack of knowledge, a reasonable balancing of opportunities versus risks is not possible. Citizens were concerned about the transparency of communication, credibility and trust in companies that bring 'nanoproducts' into the market. They refused the application of nanoparticles in the food sector. In general, every manipulation and deviation from natural growth was met with scepticism and even suspicion.

Moreover, the participants stated that *research* on MPNs and their risks should be organized and performed by international, independent authorities, by universities, or state-run institutions. They voted for an increase of funding for safety research.

Concerning *ethical and social aspects*, the participants were worried about the expensiveness of nanotechnology and thus limiting access for those who could benefit the most (unequal access), widening the divide between the industrialized and the developing world. Concerning privacy issues, they stated that the collection of increasingly sensitive data in medical diagnostics is likely to raise serious questions about information provenance and distribution, and that convergence with information and communication technology could result in possible threats to civil liberties from increasingly advanced surveillance capabilities, enabled by nanotechnologies. Moreover, the participants were concerned about subsequent developments that may be as much in the hands of users as the innovators and could be used in ways not originally intended. The complexity of the product life-cycle of nanotechnology applications may make it difficult to establish a causal relationship between actions of a company and any resulting impact. Thus, questions about sufficient liability frameworks were raised.

Regarding *regulation and control* issues, the participants were concerned whether existing regulatory regimes are robust enough to deal with nanomaterials, or whether new regulation is required. The right balance between a responsible development and safe use of nanomaterials were important for them. Like other emerging technologies that are closely linked to basic scientific research, nanotechnology generates intellectual property that is perceived as valuable and thus should be protected by patents. There is an obvious trade-off between the various laws, regulations, and treaties that govern the relationship between the public good and the protection offered by patents, they felt. The most important measure suggested by the participants in focus groups was the labelling of 'nanoproducts', which serves as a basis for deliberation and choice as well as to obtain additional information on their use, risk and appropriate disposal. But they also agreed that the consumer needs information ahead of a purchase decision: information about the (potentially) hazardous nature of a nano-ingredient enables the consumer to interpret the label and allows a risk-benefit consideration. Several participants were worried about the safety of consumer products and the lack of concrete regulations. Few citizens explicitly demanded a definitive ban (moratorium) of all 'nanoproducts'. Other participants thought of the possibility to subject 'nanoproducts' to a (governmental) authorisation after they were proven to be harmless. They concluded that an authorisation process and the obligation of long term studies would make a moratorium unnecessary.

The various aspects of concerns and perceptions found in our analysis of the outcomes of qualitative methods support, deepen, and refine the findings of quantitative surveys (like in the Eurobarometer survey), especially with regard to the possible harm to health and the environment and safety aspects. Further concerns deal with the trustworthiness and credibility of information and measures and desired communication requests. In connection with quantitative results they allow for an improved assessment of the concerns – and their basis – within the general public. They also support the findings of Gavelin et al. (2007) who analysed and discussed the results of dialogue projects dealing with nanotechnology in general, like *Nanologue* or *SmallTalk*. Gaining and maintaining public trust under conditions of scientific uncertainty seems to be the key element of the debate on perception and acceptance of nanomaterials. Openness and transparency are factors that have proven to be helpful in achieving this objective. Gavelin et al. found that the general public supports nanotechnologies that are linked to a wider social good and that it is concerned about known and unknown risks as well as the ability of the government and private sector to manage those risks. The public calls for more open decision-making about nanotechnologies. Risk communication strategies should enable a two-way communication. A transparent discussion should make available to the public informed opinions about scientific aspects, including risks and benefits, provide clear and transparent descriptions of the regulatory and funding approaches, furnish information on who has the responsibility to regulate and support nanotechnology (Gavelin et al. 2007).

5.3.4 Positions and Concerns Expressed by Stakeholders

Besides the necessity of taking into account public perception and social concerns also the interests and concerns of organized stakeholders have to be considered. The various stakeholders that have taken a position in the negotiation around 'nano' could be divided into the main groups of civil society organisations (CSO), industry and academia. CSO themselves include consumer groups, trade unions and environmental groups. In publications, stakeholder dialogues and presentations pick up

Table 5.1 Summary of the most prominent positions of different stakeholder groups on the main issues in the 'nanodebate'

CSO	Academia	Industry
Call for an increase of safety research and (partial) moratorium for the marketing of certain products.	Call for an increase of research funding.	Development of risk assessment approaches and safe handling guidelines.
Call for mandatory measures including a general labelling obligation and a harmonized traceability system. Some even call for a (temporary) moratorium.	Support for definition that is based on a defined narrow size scope with conditional exceptions (inclusion of aggregates and agglomerates).	
Call for a broader scoped definition with regard to size, also including aggregates and agglomerates.		Support voluntary measures like codes of conduct and guidelines for safe handling. Case by case decisions and assessment by scientific agencies that consider e.g. application conditions may be appropriate instruments.
Foster dialogues involving all stakeholders for equity of decision-making and public participation.	Support dialogues involving all stakeholders.	Foster stakeholder dialogues – but public 'participation' only with an informative character.

the main concerns expressed by members of the groups they represent, and cluster the various aspects. They formulate requests and recommendations for further handling of risk and improvement of governance procedures, considering the concerns raised. The main focus of stakeholder dialogues is the risk governance of engineered nanomaterials including their regulation. For consumer products, there is most discussion on nano-ingredients in food, cosmetics and other household products. Labelling and transparency are core issues next to safety. Stakeholder involvements tend to be on an invitation based on expertise and representativity (ObservatoryNano 2012). In Table 5.1, we have attempted to summarize the positions of the three stakeholder groups.

5.4 Contributions of Concern Assessment to Risk Governance

Finally, the question remains how the results of concern assessment, which bear controversies and potential conflicts, could be intertwined with the procedures of political decision-making and risk governance.

The first step is to choose a suitable and adequate method and to interpret the outcome and the gathered data carefully and with caution.

Large surveys – like this Eurobarometer study – usually ask about statements regarding nanotechnology in general. It remains unclear to which part of the multi-faceted concept of nanotechnology the respondents in these surveys refer and how these answers can be related to the more specific perceptions and concerns with regard to MPNs. On the other side, research using qualitative methods shows that most laypeople do not clearly discriminate between nanotechnology and nanomaterials. More often than not, they link risks of nanotechnology to the application of nanomaterials in various products and areas, therefore 'nanotechnology' could be read as a synonym for 'nanomaterials' within this context. In general, quantitative methods ask for existing and already formed opinions and attitudes, which are simply considered as static cognitive entities already located in individuals. Depending on personal priorities with respect to each application the responses to general questions vary from survey to survey and it is still unclear how attitudes on a single application influence the appraisal of nanotechnologies in general (Renn and Grobe 2010). In addition the responses vary with the concrete wording of the questions. For example defining the subject can prime the survey respondents.

Interpreting the results of qualitative concern assessment methods – such as focus group discussions – is also challenged by a complex system of dependent and influencing factors like concerns, perceptions, trust, acceptability, attitudes and opinions. Furthermore, some of these factors like attitudes and opinions tend to be fragile and volatile. The participants' statements about the acceptability of 'nanoproducts' in the focus group discussions indicated that attitudes and acceptance are difficult to be achieved and depend on the individual case. Experience shows that it would be overly optimistic to expect people to report insightfully on what is truly important to them or to assess what the factors are that influence their judgments and decisions. In addition, most of the qualitative methods are dynamic processes and events. People's talks, the exchange of information, hearing the perceptions and expectations of others led some participants to rethink their initially voiced positions, to formulate other, alternative statements and expectations. It became obvious to the observers that opinions cannot simply be considered as static cognitive entities already formed and located in individuals waiting to be 'excavated' by smart moderators. Many of them were formed no earlier than in the processes of interaction with other participants during the discussion (Myers 2005). This may be part of the explanation why the outcomes of focus groups on nanotechnologies have, so far, led to divergent findings. The drawbacks are based on answers, sensibilities and interactions at a point in time and challenge the interpretation of the results (Berube et al. 2011).

Thus, methods for concern assessment, especially qualitative methods, can provide no more – and no less – than first insights into people's perceptions, conceptualizations, associations and expectations regarding future technologies. With regard to risk governance their value is informational rather than instrumental. Their results can broaden the perspective of the various actors, but they do not allow for a simple delineation of governance strategies. In spite of these limitations, a number of observations can be considered for further risk governance issues (Fleischer et al. 2012b).

One example is the diffuse terminology of 'nano', which could not only be observed among laymen, but can also be consistently found among scientists, science communicators and regulators. Participants in focus group discussions used a number of metaphors in trying to better understand the content and the implications of nanotechnology (Davies 2011). In addition, a number of participants used mental short cuts or heuristics to conceptualize the unknown 'nano' terminology. For example they connected it to the more familiar natural-artificial dichotomy. When participants interpreted 'nano' as something new, artificial, they were much more sceptical about the implications of its production and use. Heuristics serve as a kind of information filter and are influenced for example by affect, values and beliefs. These kinds of 'qualitative' judgments differ from expert judgements using empirical assessment data, logical rules or probability aspects. Thus, they might shape discussion processes and their outcome differently from expert discussions. This finding may support the observation that the public's struggle with understanding science and new technologies does not necessarily emerge from a science knowledge deficit or a lack of technological literacy but rather emanates from existing personal belief predispositions and value systems that new technologies may change (Berube et al. 2011).

The second step for including concern assessment into the entire risk governance process is the 'translation' of the results into recommendations and concrete measures.

For many laymen, the concept of risk was related to the context of the application of nanoparticles rather than to the nanoparticles themselves. While people were sceptical about using nanoparticles in food, they were less critical about using them in controlled industrial environments and articulated a certain hope for using them in medicine as tools for new therapies. This could also mean that risk governance should not predominantly address nanoparticles as such but their application in different contexts – both regulatory approaches and risk communication have to address the context dependence of risk perception more specifically.

The participants of the focus group event within the STOA-project deliberated, discussed and assessed various regulatory instruments that the organizers considered to be potentially useful for risk governance of nanomaterials. Most of them were not discussed separately but rather as a combination of different measures that complement one another. In the discussion, broad consensus developed that labelling of products containing nanoparticles serves as a basis for deliberation and choice. Many participants appeared to have a 'coupled expectation' on risk governance actors. They saw the government and the consumer organizations in the role to oversee developments in products containing nanoparticles. At the same time, participants expected to be sufficiently informed about the potential risks of nanoparticles and about products containing them in order to make informed choices. Although or even because the participants in focus group discussions knew very little about nanotechnology, their claims became very clear. Independently from their different attitudes, their request as citizens and customers for clear and unbiased information coming from actors involved in risk governance was common. There was almost no trust in research results from industry. One might argue that oversight and building trust on the one hand and information and labelling on the other, are not alternatives, but merely complementary strategies for risk governance (Fleischer et al. 2012a, b).

A variety of participation events attempted to develop new forms of direct democracy in decision-making on science and technology. These upstream or midstream public engagement projects had only temporary success, but had so far not led to noticeable change and impacts on decision-making. On the other hand, public engagement in priority-setting on funding projects is widely regarded as a successful application (ObservatoryNano 2012).

In addition to public participation, stakeholder dialogues too can offer knowledge that is valuable for assessing risks and the possible approaches to managing them. But it is important that it is not the task of stakeholders at the appraisal stage to deal with normative questions like the tolerability of the risk or risk management options. By gathering information on the potential for public scepticism or social conflict in addition to experiential and practical stakeholder knowledge, concern assessment could help to identify social impacts and distinguish areas that require a more detailed analysis. In this context, concern assessment is an important part of an explicit interdisciplinary process of knowledge production.

The inclusion of scientific risk-based assessment and concern assessment in one framework could be seen as a new paradigm in the debate about the roles of sound science and precaution in decision making. By building into conventional risk analysis soft issues such as societal values, concerns and risk perceptions, as well as by looking into the interactions between various actors, such an integrated framework can lead to a better-balanced risk governance.

References

Aschberger, K., C. Micheletti, B. Sokull-Klüttgen, and F.M. Christensen. 2011. Analysis of currently available data for characterising the risk of engineered nanomaterials to the environment and human health – Lessons learned from four case studies. *Environment International* 37(6): 1143–1156.

Berube, D.M., C.L. Cummings, J.H. Frith, A.R. Binder, and R. Oldendick. 2011. Comparing nanoparticle risk perceptions to other known EHS risks. *Journal of Nanoparticle Research* 13(8): 3089–3099.

Bonazzi, M. 2010. *ANNEX – Communicating nanotechnology – Why, to whom, saying what and how? An action-packed roadmap towards a brand new dialogue.* European Commission, Unit 'Nano- and Converging Sciences and Technologies'. http://cordis.europa.eu/nanotechnology/src/publication_events.htm.

British Market Research Bureau (BMRB). 2004. *Nanotechnology: Views of the general public. Quantitative and qualitative research carried out as part of the nanotechnology study.* BMRB Social Research 2004. London: The Royal Society and Royal Academy of Engineering Nanotechnology Working Group.

Bundesinstitut für Risikobewertung (BfR). 2008. *Wahrnehmung der Nanotechnologie in der Bevölkerung. Repräsentativerhebung und morphologisch-psychologische Grundlagenstudie.* BfR-Wissenschaft 05/2008. Berlin: BfR.

Burello, E., and A.P. Worth. 2011. QSAR modeling of nanomaterials. *Wiley Interdisciplinary Reviews. Nanomedicine and Nanobiotechnology* 3(3): 298–306.

Davies, S.R. 2011. How we talk when we talk about nano: The future in laypeople's talk. *Futures* 43(3): 317–326.

Decker, M., and M. Ladikas (eds.). 2004. *Bridges between science, society and policy. Technology assessment – Methods and impacts.* Dordrecht: Springer.

Engineered Nanoparticles: Review of Health and Environmental Safety (ENRHES). 2010. Project report. http://www.nanowerk.com/nanotechnology/reports/reportpdf/report133.pdf. Accessed 20 Oct 2012.

Environmental Defense, DuPont. 2007. *Nano risk framework.* http://www.environmentaldefense.org/documents/6496_Nano%20Risk%20Framework.pdf. Accessed 20 Oct 2012.

Eurobarometer. 2010. *Eurobarometer wave 73.1.* Special Eurobarometer 341. Biotechnology. Bruxelles: European Commission.

Fleischer, T., and C. Quendt. 2007. *'Unsichtbar und unendlich'. Bürgerperspektiven auf Nanopartikel. Ergebnisse zweier Fokusgruppen-Veranstaltungen in Karlsruhe.* Wissenschaftliche Berichte FZKA 7337. Karlsruhe: Forschungszentrum Karlsruhe.

Fleischer, T., J. Jahnel, and S.B. Seitz. 2012a. *NanoSafety – Risk governance of manufactured nanoparticles* (Final report). Brussels: European Parliament, STOA.

Fleischer, T., J. Haslinger, J. Jahnel, and S.B. Seitz. 2012b. Focus group discussions inform concern assessment and support scientific policy advice for the risk governance of nanomaterials. *International Journal of Emerging Technologies and Society* 10: 79–95.

Foss Hansen, S., B.H. Larsen, S.I. Olsen, and A. Baun. 2007. Categorisation framework to aid hazard identification of nanomaterials. *Nanotoxicology* 1(3): 243–250.

Gaskell, G., S. Stares, A. Allansdottir, N. Allum, P. Castro, Y. Esmer, and et al. 2010. *Europeans and biotechnology in 2010: Winds of change? A report to the European Commission's Directorate-General for Research.* Luxembourg: Publications Office of the European Union.

Gavelin, K., R. Wilson, and R. Doubleday. 2007. *Democratic technologies? The final report of the Nanotechnology Engagement Group (NEG).* Involve: London.

Henry, T.B., F.M. Menn, J.T. Fleming, J. Wilgus, R.N. Compton, and G.S. Sayler. 2007. Attributing effects of aqueous C60 nano-aggregates to tetrahydrofuran decomposition products in larval zebrafish by assessment of gene expression. *Environmental Health Perspectives* 115(7): 1059–1065.

Hullmann, A. 2008. *European activities in the field of ethical, legal and social aspects (ELSA) and governance of nanotechnology.* European Commission. ftp://ftp.cordis.europa.eu/pub/nanotechnology/docs/elsa_governance_nano.pdf. Accessed 25 Oct 2012.

International Risk Governance Council (IRGC). 2005. *White paper no. 1 on risk governance: Towards an integrative approach.* Geneva: International Risk Governance Council.

International Risk Governance Council (IRGC). 2006. *White paper no. 2 on nanotechnology risk governance.* Geneva: International Risk Governance Council.

Krug, H.F., and P. Wick. 2011. Nanotoxicology: An interdisciplinary challenge. *Angewandte Chemie International Edition* 50(6): 1260–1278.

Leach, R.K., R. Boyd, T. Burke, H.-U. Danzebrink, K. Dirscherl, T. Dziomba, M. Gee, L. Koenders, V. Morazzani, A. Pidduck, et al. 2011. The European nanometrology landscape. *Nanotechnology* 22(6): 062001.

Marquis, B.J., S.A. Love, K.L. Braun, and C.L. Haynes. 2009. Analytical methods to assess nanoparticle toxicity. *Analyst* 134(3): 425–439.

Maynard, A.D., R.J. Aitken, T. Butz, V. Colvin, K. Donaldson, G. Oberdörster, M.A. Philbert, J. Ryan, A. Seaton, and V. Stone. 2006. Safe handling of nanotechnology. *Nature* 444(7117): 267–269.

Myers, G. 2005. Applied linguistics and institutions of opinion. *Applied Linguistics* 26(4): 527–544.

Oberdörster, G. 2010a. Safety assessment for nanotechnology and nanomedicine: Concepts of nanotoxicology. *Journal of Internal Medicine* 267(1): 89–105.

Oberdörster, G. 2010b. *Concepts of nanotoxicology.* NanoAgri 2010 conference. http://www.nanoagri2010.com/fao_mini_papers_extra_files.pdf. Accessed 20 Oct 2012.

Oberdörster, G., E. Oberdörster, and J. Oberdörster. 2005. Nanotoxicology: An emerging discipline evolving from studies of ultrafine particles. *Environmental Health Perspectives* 113(7): 823–839.

ObservatoryNano. 2012 Annual report 4 on ethical and societal aspects. In *Communicating nanoethics*, ed. I. Malsch, A. Grinbaum, V. Bontems, and A.M.F. Anderson. Maerz 2012. Brussels: EU.

Organization for Economic Cooperation and Development, Environment Directorate. 2003. *Description of selected key generic terms used in chemical/hazard assessment*, OECD series on testing and assessment number 44. ENV/JM/MONO(2003)15. Paris: OECD.

Puzyn, T., B. Rasulev, A. Gajewicz, X. Hu, T.P. Dasari, A. Michalkova, et al. 2011. Using nano-QSAR to predict the cytotoxicity of metal oxide nanoparticles. *Nature Nanotechnology* 6(3): 175–178.

Renn, O. 2008. White paper on risk governance: Toward an integrative framework. In *Global risk governance: Concept and practice using the IRGC framework*, ed. O. Renn and K. Walker, 3–73. Dordrecht: Springer.

Renn, O., and A. Grobe. 2010. Risk governance in the field of nanotechnologies: core challenges of an integrative approach. In *International handbook on regulating nanotechnologies*, ed. G.A. Hodge, D.M. Bowman, and A.D. Maynard, 484–507. Cheltenham: Edward Elgar.

Renn, O., and K. Walker. 2008. Lessons learned: A re-assessment of the IRGC framework on risk governance. In *Global risk governance: Concept and practice using the IRGC framework*, ed. O. Renn and K. Walker, 331–360. Dordrecht: Springer.

Risk Commission. 2003. *Ad hoc Commission on 'Revision of risk analysis procedures and structures as well as of standard setting in the field of environmental health in the Federal Republic of Germany' final report*. Berlin: Risk Commission.

Rocks, S., S. Pollard, R. Dorey, L. Levy, P. Harrison, and R. Handy. 2008. *Comparison of RA approaches for manufactured nanomaterials*. London: Defra.

Scientific Committee on Emerging and Newly Identified Health Risks (SCENIHR). 2007. *The appropriateness of the risk assessment methodology in accordance with the technical guidance documents for new and existing substances for assessing the risks of nanomaterials*. Brussels: European Commission.

Scientific Committee on Emerging and Newly Identified Health Risks (SCENIHR). 2009. *Risk assessment of products of nanotechnologies*. Brussels: European Commission.

Senjen, R., and S.F. Hansen. 2011. Towards a nanorisk appraisal framework. *Comptes Rendus Physique* 12(7): 637–647.

Tiede, K., A.B. Boxall, S.P. Tear, J. Lewis, H. David, and M. Hassellov. 2008. Detection and characterisation of engineered nanoparticles in food and the environment. *Food Additives and Contaminants: Part A, Chemistry, Analysis, Control, Exposure and Risk Assessment* 25(7): 795–821.

Xia, T., N. Li, and A.E. Nel. 2009. Potential health impact of nanoparticles. *Annual Review of Public Health* 30: 137–150.

Xia, X.R., N.A. Monteiro-Riviere, and J.E. Riviere. 2010. An index for characterisation of nanomaterials in biological systems. *Nature Nanotechnology* 5(9): 671–675.

Chapter 6
Ethics Research Committees in Reviewing Nanotechnology Clinical Trials Protocols

Viviana Daloiso and Antonio G. Spagnolo

6.1 Some Preliminary Remarks

The high hopes and the equally important incertitudes surrounding nanotechnologies require bioethics to examine new ways to address the ethical issues arising from their applications. These applications are debated between those who consider the novelty of these technologies in both techniques and ethical issues, and others who disagree on the need to have a specific bioethical rethink, considering nanotechnology novel only as far as its technical meaning, that is technology itself. In the first case we speak of *nanoethics*.

According to others, the difficulty to shape a more uniform bioethical reflection on the matter is due to the extreme but extraordinary possibilities which might stem this field, like the drexlerian molecular nanotechnology: nanoscale assemblers able to self-replicate. This idea has generated two different directions of viewing this issue that Gordijn expresses as follows: 'Optimistic visionaries predict truly utopian states of affairs. Pessimistic thinkers present all manner of apocalyptic visions. Whereas the utopian views follow from one-sidedly focusing on the potential benefits of nanotechnology, the apocalyptic perspectives result from giving exclusive attention to possible worst-case scenarios (Gordijn 2005).'

It is evident that some issues are common to many technologies, but others, instead, are specific to nanotechnologies because of the specific characteristics of these technologies. Implications on privacy (generated by the production of nanochips for medical care, for example), the ethics of research, the risk-benefit ratio and issues of distributive justice certainly belong to the so-called "old" issues, which are issues already known in bioethics, leading some authors to speak of 'old ethical wine in new technological bottles (McGinn 2010).' From this point of view, nanotechnologies do not give rise to qualitatively new ethical issues. To this extent

V. Daloiso (✉) • A.G. Spagnolo
Institute of Bioethics, Università Cattolica del Sacro Cuore, Rome, Italy
e-mail: viviana.daloiso@rm.unicatt.it

S. Arnaldi et al. (eds.), *Responsibility in Nanotechnology Development*, The International Library of Ethics, Law and Technology 13, DOI 10.1007/978-94-017-9103-8_6,
© Springer Science+Business Media Dordrecht 2014

it would be more justified to talk about 'ethical issues related to nanotechnology in society.' Bioethics already faces all these issues, according to those sharing the thesis of their non-novelty, as far as they concern technology more in general terms, like internet technology and the issue of using web services correctly.

The opinion may be shared that these 'old issues' do not represent qualitatively new issues, generated only by nanotechnologies, and that they would rather require stronger legislative policies. There is more than that, however. For instance, according to Allhoff, 'there are already risks in treatments and there is no good reason to think that the risks of nanosurgery are any higher than the risks typical of conventional medicine (Allhoff 2009).' Therefore, assessment is here part of the broader risk assessment that already exists in clinical practice.

As stated above, some issues are common to all technologies and it is plausible that nanotechnologies may, in that case, "reinvigorate those ethical issues". Let us take an example: the doctor-patient relationship and the shift towards the home-care technology. The extreme use of technical devices and products can certainly be ethically justified by their benefits: it is evident that medical competence involves more and more technology both in the case of nanotechnologies, and in many other technologies. Patients are becoming more autonomous and too much technology may reduce personal interaction with the doctors. This situation, while legitimately searching for less invasive solutions which would be more efficient for healthcare, risks progressively narrowing the purpose of medicine and also shifting responsibility to the patient. It also raises questions about the interpretation of the results of self-tests and their impact on the patient. Nanotechnology will act as an impulse for developments in this direction, contributing to strengthening such issues in favor of a more rigorous ethical discussion (Spagnolo and Daloiso 2009).

Regarding the concept of risk, it has been shown that it is difficult to calculate for nanomaterials, because the physical laws governing nanoscale are peculiar, while materials at a bigger scale have well-defined criteria for measuring, for example, their toxicity.

The problem lies in interpreting the concept of *nanoethics*: those who do not believe in the need of a specific bioethics for nanotechnologies base their belief on the fact that there are no changes of values and that nanotechnologies, therefore, do not require the introduction of new ethical principles.

Nanoethics does not aim at introducing new principles, but at providing a very specific assessment of nanotechnology, without addressing general bioethical concerns (Spagnolo and Daloiso 2009). For those sharing the need of nanoethics, nanotechnologies are so peculiar and specific in their extents that issues must be treated and evaluated differently within nanotechnologies themselves. To be more specific, 'the ethical profile of nanoparticle-based cancer treatments diverge substantially from that of nanomaterials in artificial joints, nanobioengineering of tissues, nanoengineered surfaces for brain-machine interfacing, nano-enabled information transmission, and engineered synthetic molecules and molecular systems to supplement the immune system. They have different objectives, pose different risks (and distribute them differently) […] have different social implications, and raise different concerns (e.g. regarding privacy, informed consent, or playing God)' (Sandler 2009).

Among those who share the opinion of a dedicated ethics for nanotechnology, Moor and Weckert (2004) remain in the impasse between for and against nanoethics. They are considered by the author as a 'nascent but an important concern' suggesting at the same time the inadequacy of referring to neither the *ethics-first model* nor the *ethics-last model*. In the first case, one runs the risk of misleading by disregarding most of the risks associated to applying nanotechnology because the technology itself is in its infancy. In the second case, the ethical evaluation could come too late, when damage has already been done.

Therefore it seems clear that the key elements of nanotechnologies represent the bases for a specific and detailed bioethical reflection. In our opinion, favoring nanoethics could be summarized as follows: nanotechnologies make it difficult to identify a systematic toxicological risk assessment of nanomaterials and nanoproducts. As far as exposure and toxicity are concerned, the mechanism through which nanoparticles enter the human body and the sites where they move or deposit are only partially known, while many others remain unknown. Concerning the dimensions and properties of nanoparticles, their capacity to self assembly, the impossibility to define the entity of possible damages for the human body, the difficulty on determining the distribution within the body of nanoparticles, are all features pertaining to nanotechnologies and therefore generate specific ethical issues.

The lack of a consensus regarding their novelty has consequences, not only in determining this assessment itself but also, and even more, within clinical trials that represent among the medical application of nanotechnologies one of the most attractive. Here, in fact, these technologies promise to speed up the discovery of new drugs and novel therapeutical agents.

The European Group on Ethics in Science and New Technologies (EGE) proposed a series of questions surrounding nanomedicine that should also be distinguished according to their short, medium and long term use. Furthermore, according to their specific use, applications can be considered as either therapeutic or non therapeutic. Consequently, the introduction of nanotechnologies in medicine might lead to the reformulation of the distinction between health and disease. According to the EGE (2007) the questions are the following:

- How should the dignity of people participating in nanomedicine research trials be respected?
- How can we protect the fundamental rights of citizens that may be exposed to free particles in the environment?
- How can we promote responsible use of nanomedicine, which protects both human health and the environment?
- What are the specific ethics issues such as justice, solidarity, and autonomy that have to be considered in this scientific domain

As far as clinical practice is concerned, the EGE identifies some possible difficulties in meeting the requirements regarding the confidentiality of patient data and data protection that are established by the international guidelines on human research ("since such data may be used by many different specialists"). In this context, a particular challenge is the informed consent. Here, as stated by the EGE, the gap between the possibilities of new diagnoses and the difficulty to interpret their results may have consequences on the informed consent in so far as this requires the subject of the consent to be understood. But regarding the introduction of nanomaterials or employing nanoparticles, the question is: "how is it possible to give information about future research possibilities in a rapidly developing research area and to make a realistic risk assessment in view of the many unknowns and the complexities?"

In this situation, ERCs can play a fundamental role.

6.2   Ethics Research Committees (ERCs)

Biomedical ethics research committees (ERCs) can be defined as independent bodies made up of people with various fields of expertise, including medical and scientific, as well as those who are responsible for ensuring that the biomedical research projects involving humans conform to the principles of biomedical research (Foster 1998).

The need of a formalised ethical review of biomedical research derives from the fact that experimentation is a moral experience that deeply worried people during the second half of the twentieth century. The creation of the ERCs derives from the need to prevent abuse of human beings: however, neither the Nuremberg Code nor the first version of the Helsinki Declaration in 1964 mentioned a reviewing commission. As noted by Spagnolo, in both of these documents the researchers were made responsible for the protection of the health and rights of the subjects involved in the research (Spagnolo 2004).

The formalised ethical review of research by "ethics commissions" followed two different paths: in the United States through the statutory system, providing for a codified, federal, legal establishment of these independent committees; in the European territory through a non–statutory system where guidelines instead of laws provide such indications. So unlike the American system of ethical reviews of research, controlled by federal regulation, the European system (particularly in the United Kingdom) sees the constitution of ERCs as the result of an initiative of the professional societies and not of regulations from the authorities. In the United States, the term "ethics committee" is generally used for a committee established in a hospital and it provides ethical consultation for clinical practice as a bedside support for clinical decision-making.

As far as the international bodies are concerned, the development of the ERCs also started in a non statutory-system. The first we will mention is an international document by a committee for reviewing research: the Helsinki Declaration at the World Medical Assembly, which was revised in Tokyo in 1975. It established that "the project itself and the carrying out of every phase of human experimentation has to be clearly defined in an experimental protocol that has to be subject to an independent committee specifically nominated for that purpose (Declaration of Helsinki 1975)."

Today the role of the ERCs has been envisaged by all the international guidelines concerning human experimentation and it is considered essential for this purpose. Its activities have always been dedicated to trying to guarantee that clinical and scientific research, which is itself a positive value that should be encouraged, does not turn against the subjects directly involved, who also contribute to the research in a fundamental way, and that it complies with a series of requirements regarding the judgement of its ethicality.

The ERCs are an ethical guarantee to protect the safety, integrity and rights involved in the experiments and to avoid scientific and economic abuse from occurring. They are also a legal guarantee because their reviewing activities are acknowledged in national and international rules and regulations. In fact, this kind of guarantee also has a fundamental value as well as positive and heralding repercussions for the quality of clinical research and for the quality of each experimentation.

The European Union, alongside the United States and Japan, intended to achieve this double aim in 1996 with the guidelines from the International Conference of Harmonization – Good Clinical Practice (ICH-GCP 1996). The aim was to promote the development and quality of clinical research using available resources in the best way and to guarantee the safety and protection of the rights of the subjects taking part.

Therefore the ERCs are called to perform a fundamental function, firstly, for the good of the subjects involved, secondly for the society and thirdly for biomedical progress. They are called to do this through even greater operative effectiveness.

The ethical evaluation of an experimental protocol involves providing an opinion on the rights of the subjects in terms of their physical, psychological and moral integrity. It involves providing an opinion on the principle of fairness and equal opportunities as well as the rights of the people who have access to the institute for assistance and who may suffer the consequences. It also includes providing an opinion on the right of the physician involved to carry out his main duty as a therapist.

6.3   The Aspects of the Informed Consent Within Nanotechnology Clinical Trials as a Challenge for the Ethics Research Committees

The pair "information-consent" in clinical trials is an element that cannot be forgotten and is still decisive. Let us think of the first document that made its necessity official, the Nuremberg Code: "The voluntary consent of the human subject is absolutely essential (art. 1)." This affirms at the same time the necessity to give "sufficient knowledge and comprehension of the elements of the subject matter involved as to enable him to make an understanding and enlightened decision." The latter element requires that "before the acceptance of an affirmative decision by the experimental subject, they should make him aware of the nature, duration, and purpose of the experiment; the method and means by which it is to be conducted; all inconveniences and hazards reasonable to be expected; and the effects upon his health or person which may possibly come from his participation in the experiment." The information represents a key-element within the Guidelines for Good Clinical Practice (GCP, European Commission 1991), where they underline the need to give the opportunity for the subjects participating in clinical research to become informed on the details of the trial (1.9): aim, benefits, risks of the study (1.15), information that has been made much clearer in the guidelines' recent version of 1996 (art. 4.8). It reaffirms, with greater emphasis, the need to refresh the informative schedule anytime new information decisive for the consent becomes available (EMEA 1996/2002).

The informed consent has then a special meaning with a decisive ethical and, to a lesser extent, medical-legal value: it is a voluntary agreement to participate in clinical trials based on the understanding of the objectives, risks and possible benefits of the research. Due to the nature of the research, the experimental field is characterized by much more grey-zones than clinical praxis. This is so that one could ask the subject undergoing clinical trial to expose himself to situations – both therapeutic and non-directly therapeutic for him – where the anticipation of possible risks and benefits is difficult, also for the researchers as they lack the information they need.

Nanotechnologies seem to fit well within these complexities due, above all, to the difficulty to currently assess their risk. So, the informed consent may lose the link to the effective understanding of risks associated to nanoparticles (that are, in a large part, not known due to the various properties of these particles that depend on many factors: size, shape, chemical composition and so on). It also depends on the expectation that society has invested in these revolutionary technologies, a hope that often exceeds the foreseen benefits.

6.4   Toxic Potential of Nanoparticles and Informed Consent

Many studies assessing the toxicity of nanoparticles are still ongoing: while some of them have found the cause of particle toxicity (Oberdörster 2001; Oberdörster et al. 2005a), others have shown a close relationship between particle toxicity and particle shape so that the toxicity depends on the modification of the surface (Hoshino et al. 2004).

If one can say that biological interactions and the toxicity effects of nanomaterials cannot always be proven, then when they would occur is similarly not foreseeable. Furthermore, according to the scientific literature, the adverse affects on human health that are caused by the physical-chemical properties of nanoparticles also relate to exposure, in other words to the mechanisms through which nanoparticles enter the human body (inhalation, skin contact, ingestion and others) and sites where they move to or deposit, and, as a consequence, determine their toxicity (Hoet et al. 2004; Gwinn and Vallyathan 2006; Oberdörster et al. 2005b).

The same characteristics that make the usage of nanoparticles, nanomaterials and nanoproducts so favorable, raise many doubts. First of all, the very small size of the particle raises doubts due to the fact that this particle can easily enter the human body. This would allow for the minimally invasive administration of drugs and, at the same time, would reduce the dosage of those administered. The effect would be, in fact, greater than that of the traditional administration and with less collateral effects. On the other hand, this trait may cause toxicity. As proven by many researchers, the behavior of materials is not foreseeable at the atomic level because of the reduction in size of the phenomena intrinsic to the nanometer scale, such as quantum physics and interface phenomena. It has been particularly demonstrated that at this size, particles or materials which are generally not very active, are toxic (Borm and Müller-Schulte 2006; Borm et al. 2006). This behavior seems to be related to, among other factors, the incremented surface to mass ratio: nanomaterials have a larger surface than the same mass of materials produced in a larger form. This can make materials more reactive extending the effects of toxicity (2nd International Symposium 2005; Oberdörster et al. 2005a). For example, experimental studies on animal-based research (rats), using nanomaterials, have shown that at equivalent mass doses, insoluble ultrafine particles are more potent than larger particles of similar composition causing pulmonary inflammation, tissue damage, and lung tumors (Oberdörster et al. 2005a). As noted by researchers, the consequent increased biological activity changes according to the modification of the physico-chemical structure of the particle and, for this reason it may produce positive and desired effects because it can produce an incremented antioxidant activity. It can also make cells enter the organism, avoid natural defences and move to organs and tissues allowing a targeted distribution of drugs, but, on the other hand, it can be negative and undesired: this is just the case of the greater toxicity shown by the individual properties of these materials, causing oxidative stress (free radicals, death of cells) (Scenhir 2005; Migliore et al. 2010).

Nevertheless, in this already complex situation one should remember that the effects (Shvedova et al. 2003; Kipen and Laskin 2005; Lam et al. 2004) depend on the chemical properties of nanoparticles and not on their dimensions (Oberdörster et al. 2005a, b; Whareit et al. 2003).

Although all studies on nanoparticles and nanomaterials' toxicity are increasing, they are still not sufficient and, as a consequence, it is not possible to concretely assess the risks associated to these particles. It seems that the current methods of assessment of risks, which are commonly used, may be not adequate to the characteristics of nanostructures, in particular with reference to their biological activity.

As already underlined by Resnik and Tinkle, *in vivo* clinical trials on animals will have significant limitations because of the different reactions that human and animal models exhibit to the exposure of substances or materials as well as to the different distribution, absorption and elimination of these materials by the organisms: what might be non- toxic to an animal, with a low concentration exposure, could be toxic to a human and vice versa (Resnik and Tinkle 2007).

Other important questions, that up to now have remained unanswered, relate to the long term affects, which are difficult to be predicted and monitored.

This incertitude comes to be more significant with regard to clinical trials, particularly regarding the informed consent. It is very difficult to explain to subjects participating in the trial, the meaning of the risk they could run when the risk and the way to manage it has not been yet defined. This situation may call for greater care in the informed consent (Marchant and Lindor 2012). Despite the uncertainty of this risk, one should minimize and limit it while defining benefits, taking into account that, at present, the long term effects of nanomedical products, nanotherapies (Bawa and Johnson 2008) are equally unknown. Furthermore, this already difficult situation becomes more difficult still because as has been noted (Resnik and Tinkle 2007), clinical trials using nanomaterials will be limited: *in vivo* trials on animals may show different reactions to nanomaterials than those foreseeable in humans. This is due to the different distribution that the substances may have in the organisms and due to their absorption. In this way, it will be very hard to make the qualitative and quantitative assessment of risks.

6.5   The Role of the ERCs in Reviewing Nanotechnology Clinical Trial Protocols

Taking advantage of the criteria for the institutional review board (IRB 2005), according to which one can delegate the board to determine the following:

- risks will be minimized
- risks will be reasonable in relation to expected benefits to the subjects or society
- provisions for data and safety monitoring will be adequate
- informed consent will be properly sought and documented
- selection of subjects will be equitable
- protections for vulnerable populations will be adequate and finally privacy and confidentiality will be protected before approving a clinical trial

it is possible to shape a role for Ethics Committees for human experimentation within nanotechnology clinical trials.

In the complex relationship between risks and benefits, the ERC may play a key – if not decisive – role because it is the public guarantor of the respect of the rights and the welfare of subjects while they contribute to more knowledge about human health by participating in clinical trials, defined as "any systematic study on medicinal products in human subjects" (EC, GCP 1991). The ERC has the ethical responsibility to verify the ethical justifiability and the validity of the information held in the information schedule or the completeness of the information itself and in the acquirement of the informed consent, as stated by the GCP. In this perspective, the ethical and scientific evaluation of an experimental protocol means a judgment with reference to the respect of human life and physical, mental and moral integrity. The document sets out in detail the way through which the ERC should verify the security, the integrity and the respects for human rights of the experimental subjects: the ability of researchers, the adequacy of structures, the rationale, the adequacy of the protocol, criteria of inclusion in the study. In particular, the ERC defined in the document as "an independent body constituted by medical professionals and non-medical members" must consider the elements listed in Table 6.1.

Table 6.1 Elements considered by ERC

The suitability of the investigator for the proposed trial in relation to his/her qualifications, experience, supporting stuff and available facilities
The suitability of the protocol in relation to the objectives of the study, its scientific efficacy, i.e. the potential of reaching sound conclusions with the smallest possible exposure of subjects, and the justification of predictable risks and inconveniences weighted against the anticipated benefits
The adequacy and completeness of the written information to be given to the subjects
The means by which initial recruitment is to be conducted and by which full information is to be given and by which consent is to be obtained
Provision for compensation/treatment in the case of injury or death of a subject if attributable to a clinical trial, and any insurance or indemnity to cover the liability of the investigator and sponsor
The extent to which investigators and subjects may be rewarded/compensated for participation

Bearing in mind that clinical trials are an important means to discover new drugs and improve tools for prevention, diagnosis and treatment, when ethically conducted, they represent not only a benefit for the person but, at the same time, they underline the importance of science. That is to say that the Ethics Committees for biomedical research must safeguard: the respect for the autonomy and the physical and mental integrity of the subjects; the scientific value of that research in the light of the progress; the equity in the allocation of the resources.

Because most of the nanotechnology research is conducted to discover drugs to cure tumors, when it comes to clinical trials the role of the ERC becomes extremely important: it must verify that the chosen methodologies are the most adequate for the aims of the protocols. For this reason, the ERC should verify the risk to be assessed in terms of probability, magnitude and duration and verify the identification in the protocol of all those elements that may influence the risk. Finally, this Committee has to make sure that any identified risk be associated to measures to prevent, minimise and monitor such a risk as much as possible: determining the levels of risk and the associated potential benefits will guarantee the protection of the subjects.

In the present search for "nanotechnology" situation, where the "first" clinical trials are under way (www.clinicaltrials.gov) or showing the first results (Davis et al. 2010), taking into account also the FDA approval of a nano-based drug in 2005, it is opportune for the ERC to give greater attention to the risk-benefit relationship to make sure that this relationship will not be intrinsically unfavorable for the trial subjects. In doing this, particular attention should be given to the informed consent and to the capacity to give a really well informed and effective consent, bearing in mind that the consent may be influenced by the seriousness of an illness.

As stated by the GCP, the object of this information regards:

- The aim of the experimentation;
- The potential benefits;
- Explanation of possible risks and inconveniences;

- Modalities of control (e.g. placebo/drugs);
- Alternatives to the proposed treatment.

The consent, therefore, should have all those elements clarifying, for example, the possibility that the active principle may exhibit different degrees of toxicity. As noted by Marchant and Lindor (2012), the informed consent process should be enhanced 'providing more information and longer consultation with prospective trials participants to ensure they understand what it is known and unknown about the nanomedicine being tested'. In this perspective it is hoped that particularly vulnerable subjects, pregnant women and children, do not take part in trials concerned with the employment of nanoproducts, or nanomaterials.

Furthermore, because too little is known about nanomaterials, and taking into account that many nanotechnologies' risks cannot be known, the ERC should pay particular attention to the too often excessive expectations placed in the so-called "nanorevolution" by the society as a whole and, above all, by the often vulnerable, patient.

Thus, the role of an ERC goes further than the close observation of the experimental protocol; the Committee takes on the responsibility towards the patient to safeguard his integrity and dignity and at the same time to give science its best chance. As the field of nanotechnology is so peculiar, it is important to have the presence, within the ERC, of an expert in this field, who can pay particular attention to the applications in medicine. Add to this, while it is clear that an expert in this field is needed, one should also remember what is stressed in the ERC statute and also stated in the WHO Operational Guidelines (WHO 4.7) that members on an ERC should continue taking part in training courses and meetings to ensure their ability to play a more complete and effective role in the protection of the subjects.

In any risk/benefit analysis, the opinion of "technical" members will be extremely important in determining how to limit or avoid a risk, the criteria for any suspension or interruption of participation of subjects, and for all aspects in which specialist/technical competence is needed.

The members with non medical and scientific expertise will be called on to pay particular attention to the ethical, legal and psychological aspects, because of the impact that the experiment may have on the subjects taking part (for example evaluating whether the participation in the experiment will excessively influence/compromise an already difficult or precarious situation caused by the pathology) as well as on the community concerned.

ERCs are particularly concerned with determining whether: risks will be minimized and will be reasonable/proportionate in relation to expected benefits to the subjects or society; provisions for data and safety monitoring will be adequate; informed consent will be properly sought and documented; selection of subjects will be equitable; protections for vulnerable populations will be adequate and; privacy and confidentiality will be protected before approving a clinical trial. Below are the elements (WHO Operational Guideline 2000) (cf. Table 6.2) to consider for the ethical evaluation of a study. The cited points are those of greatest importance in nanotechnology clinical trials:

Table 6.2 WHO operational guidelines (main points)

The scientific design and conduct of the study	The appropriateness of the study design in relation to the objectives of the study, the statistical methodology and the potential for reaching sound conclusions with the smallest number of research participants.
	The justification of predictable risks and inconveniences weighted against the anticipated benefits for the research participants and the concerned communities.
	The justification for the use of control arms.
	Criteria for prematurely withdrawing research participants.
	Criteria for suspending or terminating the research as a whole.
	The adequacy of the site including the supporting staff, available facilities, and emergency procedures.
	The medical care to be provided to research participants during and after the course of the research.
	The criteria for extended access to, the emergency use of, and/or the compassionate use of study products.
	Clear justification for the intention to include in the research individuals who cannot consent, and a full account of the arrangements for obtaining consent or authorization for the participation of such individuals.
	The insurance and indemnity arrangements.
	The adequacy, completeness, and understandability of written and oral information to be given to the research subjects.
Informed consent process	A full description of the process for obtaining informed consent, including the identification of those responsible for obtaining consent.
	Assurances that research participants will receive information that becomes available during the course of the research.
Care and protection of research participants	The suitability of the investigator's qualifications and experience for the proposed study.
	Any plans to withdraw or withhold standard therapies for the purpose of the research, and the justification for such action.
Community considerations	The impact and the relevance of the research on the local community and on the concerned communities from which the research participants are drawn.
	Proposed community consultation during the course of the research.
	The extent to which the research contributes to capacity building, such as the enhancement of local healthcare, research and ability to respond to public health needs.

The role of ERCs in nanotechnology clinical trials may be, decisive in the formulation of more specific operating procedures for nanomedicine. As already mentioned what cannot be ignored in any evaluation by the ERC is not only the scientific validity of the experimentation, but also the sufficient protection of the subject.

The scientific validity of ERCs decisions relies upon the responsibility of the various technical and scientific experts involved. Because of the crucial importance of their opinions and of their decisive consequences on the final outcome of the evaluation process, they must have the requisite expertise to fulfil their role adequately.

References

Agenzia Italiana del Farmaco (AIFA) *Osservatorio Nazionale per la sperimentazione clinica dei medicinali National Report N. 8, 2009*. http://oss-sper-clin.agenziafarmaco.it/index_ingl.htm. Accessed 11 Feb 2012.

Allhoff, F. 2009. The coming era of nanomedicine. *The American Journal of Bioethics* 9(10): 3–11.

Bawa, R., and S. Johnson. 2008. Emerging issues in nanomedicine and ethics. In *Nanoethics: Emerging debates*, ed. F. Allhoff and P. Lin, 207–223. Dordrecht: Springer.

Borm, P.J., and D. Müller-Schulte. 2006. Nanoparticles in drug delivery and environmental exposure: Same size, same risks? *Nanomedicine* 1(2): 235–249.

Borm, P.J., D. Robbins, S. Haubold, et al. 2006. The potential risks of nanomaterials: A review carried out for ECETOC. *Particle and Fibre Toxicology* 3(1): 11–46.

ClinicalTrials.gov. n.d. *U.S. National Institutes of Health*. http://clinicaltrials.gov/ct2/results?term =nano+and+clinical+trials. Accessed 11 Jan 2012.

Davis, M.E., J.E. Zuckerman, C.H. Choi, et al. 2010. Evidence of RNAi in humans from systematically administered siRNA via targeted nanoparticles. *Nature* 464: 1067–1070.

European Commission. 1991. *Good clinical trials on medicinal products in the European community*, Directive 91/507/EC 19 July. ev-lex.evzopa.ev.

European Medicines Agency for the Evaluation of Medicinal Products (EMEA). 1996/2002. *Note for guidance on good clinical practice – CPMP/ICH/135/95*. http://www.edctp.org/fileadmin/documents/EMEA_ICH-GCP_Guidelines_July_ 2002.pdf. Accessed 6 June 2014.

Foster, C. 1998. Research ethics committees. In *Encyclopedia of applied ethics*, ed. R. Chadwich, 845–852. London: Academic.

Gordijn, B. 2005. Nanoethics: From utopian dreams and apocalyptic nightmares towards more balanced view. *Science and Engineering Ethics* 11(4): 521–533.

Gwinn, M.R., and V. Vallyathan. 2006. Nanoparticles: Health effects – Pros and cons. *Environmental Health Perspectives* 114(12): 1818–1825.

Hoet, P.H., I. Brüske-Hohlfeld, and O.V. Salata. 2004. Nanoparticles – Known and unknown health risks. *Journal of Nanobiotechnology* 2(1). http://www.jnanobiotechnology.com/content/2/1/12. Accessed 20 Oct 2012.

Hoshino, A., K. Fujioka, T. Oku, et al. 2004. Physicochemical properties and cellular toxicity of nanocrystal quantum dots depend on their surface modification. *Nano Letters* 4(11): 2163–2169.

International conference on harmonisation of technical requirements for regulation of pharmaceuticals for human use 1996. *Tripartite guidelines for good clinical practice*. Geneva: International federation of Pharmaceutical manufacturers Association. http://www.ich.org/fileadmin/Public_Web_Site/ICH_Products/Guidelines/Efficacy/E6_R1/Step4/E6_R1__Guideline.pdf. Accessed 24 Jun 2014.

Institutional Review Board (IRB). 2005. *Criteria for IRB approval of research, 45CFR46.111*. http://www.hhs.gov/ohrp/humansubjects/guidance/45cfr46.html#46.111. Accessed 11 Mar 2012.

Kipen, H.M., and D.L. Laskin. 2005. Smaller is not always better: Nanotechnology yields nanotoxicology. *American Physiological Society American Journal of Physiology – Lung Cellular and Molecular Physiology* 289(5): L696–L697.

Lam, C.W., J.T. James, R. McCluskey, and R.L. Hunter. 2004. Pulmonary toxicity of single-wall carbon nanotubes in mice 7 and 90 days after intratracheal instillation. *Toxicological Sciences* 77(1): 126–134.

Marchant, G.E., and R.A. Lindor. 2012. Prudent precaution in clinical trials of nanomedicine. *The Journal of Law, Medicine & Ethics* 40(Winter): 831–840.

McGinn, R.E. 2010. What's different, ethically, about nanotechnology? Foundational questions and answers. *Nanoethics* 4(2): 115–128.

Migliore, L., D. Saracino, A. Bonelli, et al. 2010. Carbon nanotubes induce oxidative DNA damage in RAW 264.7 cells. *Environmental and Molecular Mutagenesis* 5(4): 294–303.

Moor, J., and J. Weckert. 2004. Nanoethics: Assessing the nanoscale from an ethical point of view. In *Discovering the nanoscale*, ed. D. Baird, A. Nordmann, and J. Schummer, 301–310. Amsterdam: IOS Press.

NIOSH - National Institute of Occupational Safety and Health. 2005. *Second international symposium on nanotechnology and occupational health*, Minneapolis, MN.

Oberdörster, G. 2001. Pulmonary effects of inhaled ultrafine particles. *International Archives of Occupational and Environmental Health* 74(1): 1–8.

Oberdörster, G., E. Oberdörster, and J. Oberdörster. 2005a. Nanotoxicology: An emerging discipline evolving from studies of ultrafine particles. *Environmental Health Perspectives* 113(7): 823–839.

Oberdörster, G., A. Maynard, K. Donaldson, et al. 2005b. Principles for characterizing the potential human health effects from exposure to nanomaterials: Elements of a screening strategy. *Particle and Fibre Toxicology* 2(1): 8–43.

Resnik, D.B., and S.S. Tinkle. 2007. Ethical issues in clinical trials involving nanomedicine. *Contemporary Clinical Trials* 28(4): 433–441.

Sandler, R. 2009. Nanomedicine and nanomedical ethics. *The American Journal of Bioethics* 9(10): 16–17.

Scientific Committee on Emerging and Newly Identified Health Risks (Scenhir). 2005. *Opinion on the appropriateness of existing methodologies to assess the potential risks associated with engineered and adventitious products of nanotechnologies*. Brussels: European Commission.

Shvedova, A.A., V. Castranova, E.R. Kisin, et al. 2003. Exposure to carbon nanotube material: Assessment of nanotube cytotoxicity using human keratinocyte cells. *Journal of Toxicology and Environmental Health Sciences* 66(20): 1909–1926.

Spagnolo, A.G. 2004. Ethics research committees: Procedures and quality of ethical review. In *Ethics of biomedical research*, ed. J. Vial Correa and E. Sgreccia, 234–257. Città del Vaticano: Libreria Editrice Vaticana.

Spagnolo, A.G., and V. Daloiso. 2009. Outlining ethical issues in nanotechnologies. *Bioethics* 23(7): 394–402.

The European Group of Ethics in Science and New Technologies (EGE). 2007. *Opinion n° 21 on the ethical aspects of nanomedicine*. http://www.ec.europa.eu/European_group_ethics/activities/docs/opinion_21_nano_en.pdf. Accessed 11 Mar 2012.

Whareit, D.B., B.R. Laurence, K.L. Reed, et al. 2003. Comparative pulmonary toxicity Assessment of single-wall carbon nanotubes in rats. *Toxicological Sciences* 77(1): 117–125.

World Health Organization (WHO). 2000. *Operational guidelines for ethics committees that review biomedical research*. Geneva: WHO.

World Medial Association (WMA). *Declaration of Helsinki – Ethical principles for medical research involving human subjects*, p. 75. www.wma.net/en/30publications/10policies/b3

Chapter 7
Governance of Nanotechnology: Engagement and Public Participation

Giuseppe Pellegrini

7.1 Introduction

Technoscientific innovations not only frequently generate uncertainty but also cause wide public debate and controversy. This is also the case in the emerging field of nanotechnology, even though it is still at an early stage of development and yet it has created conflict among experts, *decision makers* and the public. On closer inspection, this is a privileged moment in which to consider the relationship between the development of innovation, ethics and governance, given that the developmental stage of this technology does not allow for a definite characterisation of the main environmental and social issues that are connected to them. The design, production and deployment of nanotechnological innovations can therefore be studied in order to immediately activate pathways of public involvement, even on the basis of similar recent experiences, such as in the case of biotechnology. Principally, the lessons learned from the practices of participatory democracy linked to technoscientific innovations carried out in recent years (Joss and Durant 1995; Elder 1997; Rowe and Frewer 2000; Beierle and Cayford 2002; PDSB 2003) can guide the actions of listening, dialogue and decision-making in relation to the case of nanotechnology, avoiding the emergence of easy illusions about the possibility of obtaining a consensus 'at no cost' and establishing the boundaries of a possible confrontation between decision-makers, experts and citizens.

This chapter will discuss the reasons that have generated a growing interest among both academics and policy-makers on the issues of participation and governance in innovation. This will be followed by some research perspectives on two distinctive features of innovation processes for nanotechnology. Finally, the limitations and potential of deliberative approaches in some cases of public discussion on the topic of nanotechnology will be considered.

G. Pellegrini (✉)
Department of Philosophy, Sociology, Education and Applied Psychology,
University of Padova, Padova, Italy
e-mail: giuseppe.pellegrini@unipd.it

S. Arnaldi et al. (eds.), *Responsibility in Nanotechnology Development*, The International   111
Library of Ethics, Law and Technology 13, DOI 10.1007/978-94-017-9103-8_7,
© Springer Science+Business Media Dordrecht 2014

7.2   Technoscientific Innovations: Involvement and Participation

Over the last 20 years, the policy issues related to the *governance* of technoscientific innovations have attracted a good deal of interest. The reasons for this are both factual, with the acceleration of increasingly pervasive technoscientific phenomena and uncertain outcomes, as well as theoretical, with the renewed discussion of the role of science and technology in contemporary societies. This was initially started by the well-known studies of Jasanoff, Wynne, Irwin and Wynne, Ravetz and other scholars (Jasanoff 1990, 1995, 2004; Irwin and Wynne 1996; Ravetz 1996) who have questioned the values and ideals that up until a few years ago went unquestioned.

In particular, the idea that there was a clear separation between scientific and governance actions, while it highlighted the increase of scientific knowledge in the nineteenth and twentieth centuries, influenced the action of the state and produced new forms of public responsibility. Therefore, it is not possible to imagine science and politics as either two overlapping spheres or completely separated, but rather as mutually influenced and produced. Sheila Jasanoff has shown how the policy procedures and legal actions that take place in the courts are moments when scientific knowledge is produced (Jasanoff 1995). In this perspective, it is worth highlighting that technoscience and its products are strongly related to both beliefs and values and therefore affect issues of great public interest, such as to require forms of involvement and participation of the various audiences that may be interested in this or that innovation. The consequences of these considerations also led Irwin and Michael to state that the public in many cases possesses knowledge that could be very useful in the evaluation process of technoscientific innovations so much that in some cases it is difficult to distinguish between who is more experienced or qualified to provide appropriate knowledge to support appropriate decision making (Irwin and Michael 2003).

Current literature deals with technoscientific innovation from different disciplinary perspectives, with particular reference to governance and public participation. It is possible to distinguish at least two different points of view, that are often intertwined.

The first includes studies that consider the regulation of the uncertainty and the constant production of knowledge with the aim of studying the regulatory principles that try to protect rights and the public good: a type of scientific innovation containment model based on the value of protecting societies who wish to take on the role of *governance* (Stehr 2004). Scientific knowledge developed in this historical phase, according to this perspective, is very different from that developed after the First and Second World Wars. It is a set of technical sciences that would require a greater effort in monitoring and an intense regulatory control, thus highlighting the need for new ways of developing knowledge policies and research, due to their powerful ability to change the body, mind and environment.

The second perspective proposes a reflexive politics of knowledge, based on the contemporary approaches of STS studies (social studies of science and technology) developed, in particular, by Sheila Jasanoff and Helga Nowotny in order to indicate uncertain boundaries between science and society. This second interpretation focuses on the limits of a division between the knowledge systems of experts and the non-expert public, as well as the distinction between facts and values, trying to accentuate the ability to analyse and reflect on their training. Thus, rigid distinctions between subsystems must be overcome, promoting the development of a theory that valorises social diversity and highlights the interactions between science and society (Jasanoff 2004; Nowotny et al. 2001; Nowotny 2003).

Both perspectives are based on several common assumptions. The first of these assumptions refers to the fact that the recognition of the difficulty to calculate risk levels has generated confusion and disorientation in many events, displacing both policy makers and experts, so that on several occasions, they have been unable to give reassurances and clarification, resulting in a crisis of the credibility of many institutions assigned the governance of science and innovation. These are accompanied by the difficulties the *decision makers* have to deal with, who are often powerless in the face of the need to predict the consequences of specific innovations, especially in certain areas that seem to have a pervasiveness that has never been expressed before. Contemporarily, the demands of citizens have also increased, both as individuals as well as spokespersons for larger organizations, to the extent that it is necessary to generate unprecedented involvement and lead many governments to activate forms of consultation and public participation. These initiatives have increased the number of actors capable of participating in the formation of technoscience *policies*, recognizing all the limitations of traditional settings and decreeing the failure of a theory of 'technocratic proxy' in the governance of science. With the increase of these demands, the inescapable need to activate more transparent and inclusive decision-making processes is evident.

All the aforementioned reasons have generated a cultural, political and social movement, which has focused particularly on the forms of *involvement* as well as the issue of public *participation*. The aim of this paper is to discuss the two forms of governing technoscientific innovations, i.e. the set of actions and decisions that aim to ensure their proper management.[1]

At this point, it is worth trying to define what is meant by involvement and participation, stating that the two terms will be dealt with by referring to those innovations that will have a major impact on health and the environment, and therefore directly involve citizens as well as various members of society.

In relation to technoscientific innovations, *involvement* is intended as the set of actions that occur on several levels of the relationship between experts, decision makers, stakeholders and the public. These are initiatives that aim to open public spaces of discussion where to address the phenomena of innovation from multiple

---

[1] In this context, reference is made to the concept of *governance* formulated by Le Gales (1998): 'the coordination process of actors, social groups and institutions to achieve their objectives discussed and decided collectively in fragmented, uncertain environments'.

perspectives, thus valorising the role of expert knowledge along with the issues and knowledge proposed by citizens and organizations of society. These initiatives, promoted largely by public institutions, are based on the principle of inclusion in order to encourage a dialogue between the various actors by activating a comparison that allows different subjects to reconsider their own positions, overcoming ideological barriers or pre-assumptions that assign to the discoveries of experts the ability to establish themselves as objective realities capable of determining the right policy measures (Felt et al. 2007). It is possible to use the term involvement as a synonym of *consultation*: a mechanism that is configured in terms of communication as, generally, a one way relationship, even though it is not alone and mere information.

The involvement of experts, decision-makers, citizens and organizations of society can be expressed in various ways: from a simple presentation of scientists dealing with a specific innovation to the possibility of offering a wide range of proposals that highlight the impact on health and the environment. Whoever uses this type of procedure obviously does not claim to improve the decision-making processes and tends rather to emphasize the principle of transparency, providing information and identifying the responsibilities involved with opportunities for open discussion, with the possibility of comparison between actors who would not normally have other opportunities to interact (Weale 2003; Liberatore and Funtowicz 2003; Cross 2003). Generally, these actions occur when the decisions related to the adoption of a particular innovation have already been taken.

Whereas, when referring to *participation*, it should be considered not so much as a form of consultation or involvement, but rather as the possibility that a person has of being involved in particular innovation dynamics in order to be able to contribute to the articulation of the debate that is being generated around a particular techno-scientific issue. According to this perspective, it is possible that these non-expert actors may also take part in a 'co-production' process of knowledge (Nowotny 2003), actively interacting with experts and policy makers. Participating – *taking part* – therefore means including not only technical points of view, but also ethical, social and economic ones that can be used in the decision path connected to the governance of a particular technoscientific innovation. Citizens and experts can therefore be involved in forms of discussion, control and validation of certain innovations, forming so-called 'epistemic communities', i.e. those contexts for discussion and debate among social, political and economic groups, who are conveyors of different knowledge systems (Callon 2003).

What distinguishes participation from involvement is, in short, the weight of the inclusion of the different actors in processes that can affect the cognitive, ethical and decision-making aspects of the subject of the discussion. Such practices usually take place prior to the decisions on the adoption of a particular innovation and refer to the so-called deliberative democracy as a 'process based on public discussion among free and equal individuals' (Pellizzoni 2005, 14), primarily intended as a place of dialogue or discussion preceding the decision. It is possible to imagine the terms *involvement* and *participation* as two elements in a continuum ranging from highly vertical forms of consultation and listening, where some actors have more power in activating certain actions in contrast to generally passive interlocutors, to

Involvement	Participation
Public meetings	Citizen juries
Forums	Consensus conferences
Citizen panels	

Table 7.1 Involvement and participation procedures in the governance of techno-scientific innovations

more dialogic and equal forms of participation, where the recognition between the actors is very strong, beyond their role and power, with the power to change their points of view on the matter. The table below shows several forms of involvement and participation that are commonly used in various European countries in the discussion and *governance* of technoscientific innovations (Table 7.1).

The use of particular forms of involvement and participation depends on several factors that may affect the *governance* of technoscientific innovation, influencing any public choices as well as the forms of development related to them.

First of all, the type of technoscientific innovation which is the subject of debate has a specific relevance. An innovation that relates to very general problems such as the use of stem cells for medical research or biotechnology for plant transformation is different from the construction of nuclear power plants and the sale of food containing GMOs. The weight of the externality, in other words, the anticipated impact, takes on a very different significance in the eyes of citizens, and is not the same as promoting the activation of spontaneous public arenas as assemblies rather than the more selective and guided procedures of citizens' juries and *consensus conferences* on the object of interest (Bobbio 2002; Pellegrini 2005). The first is a group of local citizens that includes representatives of the communities affected by particular technoscientific innovations who make decisions, while a *consensus conference* involves the creation of a group of citizens, the interaction with experts and the establishment of a final document to be submitted to the attention of a wider group of stakeholders and citizens. The nature of an innovation can influence the type of inclusion of the actors thus promoting different types of approach: from free and informal to more focussed and structured. More critical studies of participatory procedures point to the fact that when they are activated within the context of innovations that are not very well known by the public, as in the case of nanotechnology, a forcing is realised. A co-optation mechanism which tends to favour the contribution of highly motivated citizens and social groups, and therefore not representative of the population, draws attention to issues that are not yet in the public domain, gathering the views of a limited segment with the false idea that it has actually collected the common opinion (Regonini 2005).

Another highly significant element for the development of involvement and participation relates to the *discursive structure* of the problem. This dimension includes the definition of the technoscientific innovation that is proposed by the different actors involved, with all its potential and limits, as well as the consequent power to propose a specific 'interpretive frame'. In this sense, the different 'discourses' of a specific innovation offer different interpretations and assessments of its usefulness, the various types of impact on health and the environment, with it being possible to observe the deployment of different modes of judging the economic issues of sustainability and ethical problems. All these elements are presented as critical to the governance of a technoscientific innovation. At this point, it goes without saying that the actors who are more able than others to have a material impact on the 'discourses' and 'interpretive frames' within which a specific theme is presented have an extraordinary power of articulating and of orientating the discussion. However, these principles also apply to actors who do not have any form of recognized authority from the outset of the debate, as in the case of organized citizen groups who take action to demand the study of a rare disease with experimental drugs, and therefore may fail to gain the attention of the media and the public.

The subject of *voluntary* or obligatory paths of involvement and participation should not be overlooked. The various forms of negotiation within the environmental context such as Agenda 21 have a rather high degree of formalization and regulation, with them often being harshly criticized for their rigidity as well as their symbolic nature in cases where the real decisions are taken outside disregarding the law. It is evident that the application of these tools is very different from the use of *ad hoc* forms, appropriately chosen and on the basis of certain conditions set by institutional actors or promoted by civil society organizations. When discussing forms of involvement and participation, we exclude the above-mentioned institutional forms that belong to the category of negotiation instruments and that are already structured and used with full public legitimacy.

Finally, it is also worth considering the different *phases* of the technoscientific innovation process in which the initiatives of involvement or participation will be activated in relation to the objectives to be pursued. If wanting to include stakeholders prior to the decision-making process, it is possible to use participatory procedures with less risk of seeing frustrated expectations of the participants, since their ability to affect the configuration of the discourse and decision-making will be greater. In other cases, the procedures will be dedicated to the empowerment and monitoring of certain technoscientific innovations, and therefore more oriented to consultation and involvement with a different level of power, provided it is clearly disclosed in advance.

7.3   Nanotechnologies, Participation and Other Forms of Involvement

Nanotechnologies are often classified as part of the so-called 'converging technologies', which are supposed to be integrated with other forms of techno-scientific innovation such as biotechnologies, computer science, robotics and so on. Despite holding good in terms of definition, such a degree of integration is not registered in terms of public perception since it is generally proved that citizens are able to tell the difference between different technological applications, and even within each application as the case of biotechnologies may be – e.g. see the difference between biomedical and food biotechnologies (Bucchi and Neresini 2006).

Currently, nanotechnologies are not a highly controversial topic in the public eye and the development of their application still has not required advice from an expert whose study-application context comes from outside, more generally or specifically speaking, in order to assess the risks and the opportunities available for the whole community. The level of risk and uncertainty stemming from nanotechnologies is not at the core of public debate and anxiety towards them has not been reported as a significant fact in the latest studies on public opinion (Neresini 2007).

Still, it is difficult to tell what's the difference between the different kinds of public as far as this area of techno-scientific innovation is perceived, and governments do not seem to be taking measures against the advancement of medical research, the development of new materials, or consumer goods.

Nevertheless, several programs have been set up – except in the case of biotechnologies – in order to inform, discuss and maintain contact with the public on central issues of social, economic or health interest. The European Union has been funding research projects with the aim of studying public attitudes, and it has also created a discussion forum in order to understand the trends of the different stakeholders concerned about the risks and anxieties about nanotechnologies. During 2007, a program of public consultation was launched by the European Commission along with a project aimed at defining a 'Code of Conduct' in order to promote a more responsible development of research. By means of this initiative, the member states were encouraged to take concrete action for the sake of understanding, sustainability, prevention, inclusion, excellence and responsibility (EC 2007).

Among the ongoing and already-funded projects, we suggest the research network FramingNano and the Nanoforum platform, which are facing issues such as the risks and anxieties perceived by the public from various viewpoints, more specifically the so-called EHS (environment, health and safety) and ELSI (ethic, legal and social issues).[2]

However, participation in this kind of initiatives has been restricted to selected groups of citizens so far, and the media coverage is still not as substantial as to provide widespread and updated pieces of information or commentaries. Thus, it can be deduced that nanotechnologies is currently a rather dark matter, whose public and many of its stakeholders – both as individuals and as a group – do not have enough knowledge to give their opinions.

However, there exist other key elements which are going to make the governance of nanotechnologies quite tough in the near future.

An element of crucial importance is first of all the wide variety of the materials produced in any specific area, which brings about the difficulty to identify its peculiarities and key principles. Dealing with nanotechnologies for medical research on the one hand, and for materials or energy on the other, are two different things. Moreover, we are not provided with data on the impact that these innovations can have. Even the availability of updated sources of information is rather open to debate, assuming that these are ongoing experimental researches, which have not

---

[2] See the following websites: http://www.framingnano.eu and http://www.nanoforum.eu.

been made official or patented yet. As a result, they have no readily available ad hoc sources of information built on an ad hoc basis.

Furthermore, we cannot neglect the fact that the trend expected by many research centres implies a shift from 'passive' to increasingly advanced nanotechnologies, that is to say to 'nano' applications which will bring about more and more intelligent – and somehow autonomous – forms of nanotechnology, and therefore hardly controlled in the collective imagery (Felt et al. 2014; Roco 2008).

As far as mass communication is concerned, significant resources are often unavailable and those who work on innovation in the field of nanotechnologies are provided with means and sources which are especially designed for the production and implementation of technological devices. The initiatives promoted by the European Union with some member states – along with the collaboration of various organizations that are involved in 'science and society' – are still in the embryonic stages, more focused on the issue of the risks than on the involvement coming before the decision-making process, that is to say the phase in which the interpretative framework of nanotechnologies is set (EC 2008).

Even on the legal front, there are problems of control since the authorities which are in charge are often unaware of their areas of competence, and therefore they cannot give information and answer the questions coming from the public or organizations belonging to the civil society.

Given that the very issue of nanotechnologies has not fully become part of the media debate and it has not caused significant controversies yet, one can wonder which suggestions need to be considered from the viewpoint of involvement and participation linked to these issues and, as a consequence, which governance trends can be drawn. In order to reach this goal, we suggest some possible scenarios which take account of two key elements for managing activities of involvement and participation linked to the issue of nanotechnologies – relevance and uncertainty – if we follow the interpretative model offered by (Radaelli 2002; Pellizzoni 2001).

Such consideration must be given by bearing in mind that the field of nanotechnologies cannot be examined in general terms, as was the case at the time of biotechnologies. On the contrary, the debate should be articulated in the public arena with a high degree of specificity. To this end, it is important to draw a distinction between nanotechnologies in medicine and nanotechnologies of civil, industrial, military purposes and so on. Based on what happened with biotechnologies, it can be assumed that different forms of application will produce – in the future – different levels of understanding and acceptance from the public. In this context, we refer to relevance as the strength that a certain theme of public interest has to stimulate participation, both in the stakeholders involved and in the public. What we regard as uncertainty is instead represented by the elements of supposed risk which cannot be estimated when a new technological innovation is introduced or can be included on account of supposed benefits (Renn and Roco 2006). For these two factors, we both refer to the perception of the public and to their changes in time. The Italian asbestos case is emblematic because it demonstrates that – even though the risks on employing asbestos in specific areas were already well-known – it was only after a strong public exposure of its harmful results that there was an increase in relevance so that its use was finally forbidden (Gallino 2007).

Table 7.2 Uncertainty and relevance in the governance of nanotechnologies

		High uncertainty		
Low relevance	Technocratic governance		Participatory procedures	High relevance
	Bureucratic procedures		Traditional governance	
		Low uncertainty		

Adapted from Radaelli (2002) and Pellizzoni (2001)

Depending on the kind of governance to be undertaken by public decision-makers, different scenarios are likely to be made conceivable in shaping the policies and dynamics of participation. Table 7.2 sketches out these changes by drawing a distinction between the two levels of relevance and uncertainty.

As we have already mentioned, the situation which is most similar to what is currently occurring to nanotechnologies matches the description of low relevance and low uncertainty (bottom-left box). This is a situation in which politics can be controlled bureaucratically, without special consultation or participation processes. This very situation occurred when biotechnologies were at the beginning and, to a certain extent, it seems that European organizations are trying to get nanotechnologies moving in a different direction. Their attempt is, for instance, to make space for informed debates and study the phenomenon in terms of communicative effects in order to find the possible crucial points and have certain guarantees.

To a certain extent, this situation can be seen as a way of avoiding the 'dead calm' effect which, in other cases, anticipated the explosion of bitter controversies with the aim of giving information and developing forms of communication that enable us to create a solid fact-finding basis. This can be used whether there might be clashes of opinion between the different stakeholders involved and, more generally, with the public. The purpose of this informative effort is also to avoid ideologically-driven debates where preliminary questions advanced by some stakeholders can possibly interrupt communication between the parties. Because of that, the emphasis is very much on explaining what nanotechnologies are, what they are useful for and how it is possible to distinguish between the great variety of applications. This distinction is drawn by trying to suggest a pragmatic consideration of them, along with a series of measures meant to guarantee control in terms of the effects caused on the environment and on human health.[3]

---

[3] As a link to that, see the plenty of initiatives launched by the European Union along with other countries. Among them, we can recommend 'nanoTruck', which is a journey across Germany to disclose the principles and the areas used for nanotechnologies (http://www.nanotruck.de). We also recommend 'Interactive journey into the nanocosmos', a website which has been created for showing the nanoscale dimensions through which specific applications can be shaped (http://www.nanoreisen.de). Also, the French ministry of research launched the following website: http://www.nanomonde.fr. In 2005, a brochure called 'À la découverte du nanomonde' was distributed in order to illustrate the so-called 'nanoworld' (http://www.nanomicro.recherche.gouv.fr/docs/plaq.nano-monde.pdf). In the United Kingdom, a series of computer devices has been developed in order to

If we focus on 'low relevance' and 'high uncertainty', it should be easier to achieve technocratic governance, that is to say that kind of governance led by technical organizations and officers who do not include external advice or viewpoints beyond generally accepted expertise positions in decision-making (top-left box). In this case, those who are supposed to establish mechanisms of governance tend to choose a technocratic orientation, assuming that a possible lack of interest from the public might allow us to proceed smoothly without developing special forms of consultation or participation.

As far as the extent of nanotech applications is concerned, this kind of orientation is likely to be seriously criticized in some case. For example, the use of medical innovations are likely to cause controversies between officially recognized science and patients associations, leaving aside the ethical implications that these innovations might have as well as the possible protests by organizations belonging either to the civil society and the churches of different countries.

Where the level of uncertainty and relevance are both high (top-right box), in other words in a context where the public attach great importance to a specific nanotech application, for instance, it is more convenient to promote the active participation of the stakeholders. This is meant to enhance their ideas and contributions and also to avoid closed-door decisions, which can run the risk of being refused by social organizations and communities (Petersen and Bowman 2012).

This is the typical case in which the so-called epistemic communities can be included in the participation processes by increasing their potential to develop their knowledge and responsibility. As often happens in the environmental field, these

---

improve the knowledge of nanotechnologies stemming from Oxford University (http://www.conted.ox.ac.uk/courses/professional/nanobasics/nano/interface.html). In 2005, London's Science Museum hosted the exhibition 'Nanotechnology: Small Science, Big Deal' by using a multimedia platform in order to show how nanotechnologies work (http://www.sciencemuseum.org.uk/antenna/nano/index.asp). All these initiatives became part of a public information campaign which did not arouse great controversies. What attempted to highlight the potential risks and anxieties of the public was a report published in 2004 by the Royal Society with the title 'Nanoscience and Nanotechnologies: Opportunities and Uncertainties' (http://www.nanotec.org.uk/finalReport.htm). Another attempt to collect ideas from the public is given by the 'Nanotechnology Engagement Group', which was established with the aim of promoting public involvement through projects such as 'Nanodialogues – Experiments in public engagement with science'. These projects were supported by Demos and Lancaster University in 2005. There, the public had the opportunity to talk with the scientists about governance, research funding and other issues linked to the development of nanotechnologies (http://www.demos.co.uk/projects/thenanodialogues/overview).

Another initiative which had a strong public impact was NanoJury, carried out in 2005 along with Greenpeace, 'the Guardian' and some UK universities. A group of 25 randomly selected citizens joined in a debate which produced a paper of advice on health, social as well as environmental issues and their normative requests to be addressed to decision makers. The 'Code of Conduct' is an initiative by the Royal Society whose aim is to promote – along with the Nanotechnology Industries Association (NIA) – a responsible development of nanotechnologies. A similar initiative was promoted by a study group as for the shaping of the 'NanoCode', with the aim of getting involved in the debate on the technical, social as well as commercial issues linked to nanotech innovation. Published in 2008, the code suggested the seven best practices to be followed by the organizations on a voluntary basis (http://www.responsiblenanocode.org).

communities can be made up of professionals who are able to exert a strong influence on decision-makers thanks to their generally accepted authority.[4]

Moving on to situations in which high uncertainty and low relevance can occur, innovation initiatives of traditional and bureaucratic governance are likely to be more frequent (bottom-right box). In this field, the lack of awareness among citizens and their organizations is due to the cognitive as well as existential abstraction of the topic, for which there is no need for inclusion in the public debate or ad hoc forms of involvement.

Clearly, this typology is approximate and puts forward models which are subject to alterations and mutual contaminations since the governance of any innovation can take different directions and the stakeholders that contribute to and have influence on these processes can often choose whether to adopt other typologies of involvement and participation even if the latter are not considered by public authorities.

7.4   Nanotechnologies and Deliberative Processes

As mentioned above, a scenario in which there is a high level of relevance and uncertainty is more suitable for introducing processes of inclusion and participation, which is to say spaces of deliberative democracy where the stakeholders involved meet in order to advance their position on discussing in ad hoc contexts.

These are procedures through which we can acquire greater understanding as long as communication is handled effectively, that is to say by ensuring transparency in discussion processes among independent subjects. In this sense, we can refer to the phrase 'deliberative model' as a discussion between different stakeholders who are likely to reconsider their original position on the theme discussed. If that is the meaning, 'deliberative' is synonymous with 'debated confrontation', it is not only a question of decision (Pellegrini 2008). For those who stand for the deliberative model, it is the outcome of the discussion which can produce better decisions. It is important to point out that these two moments – discussion and decision – are significantly different. As a matter of fact, the more deliberative practices are complementary to those mechanisms of representative democracy, the more they are effective. As a link to that, it is worth remembering that the role played by the elected representatives is the most relevant to decision-making procedures since the democratic governments in which they have been tested are representative in kind. In this sense, public decision-makers cannot abandon their role although in deliberative contexts they are asked to perform transparent and inclusive decision-making procedures (Pidgeon and Corner 2013).

---

[4] Adler, E., and Haas, P. (1992), Epistemic communities, world order and the creation of a reflective research program, *International Organization,* 46(1): 367–390.

Deliberative practices help improve the debate by enabling the parties to follow its progression carefully. As a result, discussion and inclusion are key element in terms of deliberative process. The hypothesis which underlies these procedures is the result of an honest attitude which is able to produce 'better decisions' (Eeten 2001), but they are not necessarily 'the best' over all.

On studying the results of a recent consensus conference attended in the United States on the very issue of nanotechnologies,[5] we can detect some of the effects caused by the deliberative model and point out its theme connections along with the emphasis on the issues under debate.

In the first place, citizens' suggestions stem from the need to define the term nanotechnologies correctly. This forms a sort of basic condition which comes before legal constraints and other guarantees provided for the protection of human health and the environment. As far as communication is concerned, it is necessary to make wider space for nanotechnologies in the mainstream media as well as the most popular TV channels, provide relevant and clear pieces of information from the experts and give free access to the test results on the materials employed by means of nanotechnologies.

Among the most interesting points, the organizers have highlighted an increase in awareness in the attendees, who managed to acquire a wider knowledge of the issue thanks to the deliberative model as well as the process of empowerment which put them in contact with stakeholders and decision makers in a new way (Powell and Kleinman 2008). On the other hand, some critical points cannot be underestimated, such as the cynical attitude towards the possible effects of citizens' suggestions on decision-making choices in nanotechnologies. Such skepticism has also been underlined in the literature of the field (Guston 1999; Einsiedel and Eastlick 2000; Gastil and Dillard 1999; Kleinman 2000).

7.5   Conclusions

The case of nanotechnologies presents again a well-known issue in the field of techno-scientific innovations: the debated legitimacy of techno-science as a safe and never-falling device to handle innovation and render public opinion properly.

Politicians and experts are put through the mill in framing a valid up-to-date image of the public and civil society, and therefore avoid taking for granted assumptions or preconceptions about them.

---

[5] This is the report resulting from a consensus conference which took place in the United States in 2005 (Kleinmann and Powell 2005). The meeting involved a group of 13 citizens and 7 experts and it adopted a participatory procedure throughout three weekends. After discussing with the experts, the citizens filed a final document in which they gave their suggestions on issues such as the environment, human health, control, the media and public participation. All these were connected with the future development of research on nanotechnologies. The initiative was supported by the Nanoscale Science and Engineering Center from the University of Wisconsin and the UW Integrated Liberal Studies Programme, Wisconsin.

These kinds of preconceptions are used, for instance, when it is believed that expert knowledge is undoubtedly useful for society and it can always be used correctly (Beck 1992). The result is then a linear model of communication which stems from the experts and is addressed to politics and society, assuming that the latter has to accept pieces of information uncritically and passively. In this regard, it must be remembered that in the latest 10 years the degree of awareness about and interest in health and environmental issues has increased and the possibility that citizens have to contribute to the debate on innovation is an essential element. Starting from this assumption, it is possible to list some crucial factors which are to be considered peculiar for the governance of these innovations.

First of all, we do not know the 'best' way to handle involvement/participation procedures. Depending on the situations and the innovations brought into play, we can perform ad hoc forms of involvement and participation by considering the conditions, the available sources and the kind of democratic tradition on which they are based. Clearly, the deliberative model is not useful where there are any controversies or clashing conflicts between the parties as has happened with high-speed rail lines. As a link to that, it must be underlined that common forms of negotiation can be used – without undertaking untracked paths – so that naïve positions and weak forms of participations can be overcome.

Another misunderstanding to be avoided whenever we intend to handle governance processes through forms of involvement and participation is represented by the idea of avoiding conflicts. Often politicians, but also experts and stakeholders, have the bad habit to regard consensus as a goal to be achieved, always in every case. In a period in which there are multiple ways of approaching progress and welfare, techno-scientific innovations cannot be achieved through generally agreed consensus. At this point, it is important to mark the boundaries between success and failure, which occurs when the forms of involvement and participation are enabled. Making the public and different stakeholders become involved actually means trying to improve the quality of decision-making not only by facing controversies openly, but also by putting pressure on the different layers of responsibility which everyone has to respond to.

If the mechanisms of involvement are developed, these kinds of situations can be less influential upon decision processes, but there might be conflicts which make the debate more complex.

Anyway, if we want to follow a process of governance which pays more attention to public involvement and participation, the role played by experts must be reconsidered. Given the public nature of today's techno-science, the role played by the experts will have to be shaped by means of a closer confrontation with the issue of innovations, in a framework of re-established relationships.

Arguably, this does not mean that researchers and scientists are supposed to act as communicators or cultural mediators; scientific institutions should consider certain dynamics of communication which are included in the processes of involvement and participation, even if they are not carried out by researchers. On assessing the possible risks, it is crucial to have an early involvement and confrontation between the different stakeholders involved in nanotech innovation (Renn and Roco

2006). The experiences gained in the latest years have tested different means of involvement and participation which might be used in the different phases of application and implementation (Rowe and Frewer 2000; OECD 2002). Enabling the experts to come out into the open during the phases of pre-assessment is one of the most urgent challenges in nanotechnology.

The ambiguities and uncertainties implied in the development of nanotechnologies will possibly be faced with appropriate procedures of involvement and participation in the coming years. Thus, the possible scenarios will be complex and they will require an ability to interpret the interactions between citizens, experts and stakeholders by taking account of both techno-scientific aspects and values such as responsibility and power. The biggest challenge in this field is the ability to handle these processes by supporting not only the governance of these innovations, but more so a new democratic way of facing issues which have a strong public impact.

References

Beck, U. 1992. *Risk society*. Cambridge: Polity Press.

Beierle, T.C., and J. Cayford. 2002. *Democracy in practice: Public participation in environmental decisions*. Washington, DC: Resources for the Future.

Bobbio, L. 2002. Le arene deliberative. *Rivista Italiana di Politiche Pubbliche* 3: 5–29.

Bucchi, M., and F. Neresini (eds.). 2006. *Cellule e cittadini. Biotecnologie nello spazio pubblico*. Milano: Sironi Editore.

Callon, M. 2003. The increasing involvement of concerned groups in R&D policies: What lessons for public powers? In *Science and innovation*, ed. A. Geuna, A.J. Salter, and W.E. Steinmuller, 30–68. Cheltenham: Edward Elgar.

Cross, A. 2003. Drawing up guidelines for the collection and use of expert advice: The experience of the European Commission. *Science and Public Policy* 30(3): 189–192.

Eeten, M.V. 2001. The challenge ahead for deliberative democracy: In reply to Weale. *Science and Public Policy* 28(6): 423–426.

Einsiedel, E., and D. Eastlick. 2000. Consensus conferences as deliberative democracy. *Science Communication* 21(4): 323–343.

Elder, M.J. 1997. The process of community involvement. A case study: The Bartlesville, Oklahoma, lead project toxicology and industrial health. *Toxicology and Industrial Health* 13(2/3): 395–400.

European Commission. 2007. *Governance and ethics of nanotechnology*. http://ec.europa.eu/research/science-society/index.cfm?fuseaction=public.topic&id=1524. Accessed 24 Jun 2014.

European Commission. 2008. *Third international dialogue on responsible research and development of nanotechnology*. http://cordis.europa.eu/nanotechnology/src/intldialogue.htm. Accessed 20 Oct 2012.

Felt, U., et al. 2007. *Taking European knowledge society seriously*. Brussels: Office for Official Publications of the European Communities.

Felt, U., S. Schuman, C. Schwarz, and M. Strassnig. 2014. Technology of imagination: A card-based public engagement method for debating emerging technologies. *Qualitative Research* 14(2): 233–251.

Gallino, L. 2007. *Tecnologia e Democrazia*. Torino: Einaudi.

Gastil, J., and J.P. Dillard. 1999. Increasing political sophistication through public deliberation. *Political Communication* 16(1): 3–23.

Guston, D. 1999. 'Evaluating the first US consensus conference: The impact of the citizens' panel on telecommunications and the future of democracy. *Science, Technology, and Human Values* 24(4): 451–482.

Irwin, A., and M. Michael. 2003. *Science, theory and public knowledge.* Oxford: Oxford University Press.

Irwin, A., and B. Wynne (eds.). 1996. *Misunderstanding science?* Cambridge: Cambridge University Press.

Jasanoff, S. 1990. *The fifth branch. Science advisers as policymakers.* Cambridge, MA: Harvard University Press.

Jasanoff, S. 1995. *Science at the bar: Law, science and technology in America.* Cambridge, MA: Harvard University Press.

Jasanoff, S. 2004. Science and citizenship: A new synergy. *Science and Public Policy* 31(2): 90–94.

Joss, S., and J. Durant. 1995. The UK national consensus conference on plant biotechnology. *Public Understanding of Science* 4(2): 195–204.

Kleinman, D.L. (ed.). 2000. *Science, technology, and democracy.* New York: State University of New York Press.

Kleinman, D.L., and M. Powell. 2005. *Report of the Madison area citizen consensus conference on nanotechnology* (2005, April 25). Nanoscale Science and Engineering Center at the University of Wisconsin.

Le Gales, P. 1998. La nuova political economy di città e regioni. *Stato e mercato* 53(1): 53–91.

Liberatore, A., and S. Funtowicz. 2003. Democratising expertise, expertising democracy: What does this mean, and why bother? *Science and Public Policy* 30(3): 146–150.

Neresini, F. 2007. Prima della prima. *Sapere* 73(4): 6–13.

Nowotny, H. 2003. Democratising expertise and socially robust knowledge. *Science and Public Policy* 30(3): 151–156.

Nowotny, H., P. Scott, and M. Gibbons (eds.). 2001. *Re-thinking science: Knowledge and the public in an age of uncertainty.* Cambridge: Polity Press.

Organization for Economic Cooperation and Development (OECD). 2002. *Guidance document on risk communication for chemical risk management.* Paris: OECD.

Pellegrini, G. 2005. *Biotecnologie e Cittadinanza.* Padova: Gregoriana Editore.

Pellegrini, G. (a cura di). 2008. *Technoscientific innovation and new forms of democracy.* Soveria Mannelli: Rubbettino.

Pellizzoni, L. 2001. Democracy and the governance of uncertainty. The case of agricultural gene technologies. *Journal of Hazardous Materials* 86(1–3): 205–222.

Pellizzoni, L. (a cura di). 2005. *La deliberazione pubblica.* Roma: Meltemi Editore.

Petersen, A., and D. Bowman. 2012. Engaging whom and for what ends? Australian stakeholders' constructions of public engagement in relation to nanotechnologies. *Ethics in Science and Environmental Policies* 12: 67–79.

Pidgeon, N., and A. Corner. 2013. Nanotechnologies and upstream public engagement. Dilemmas, debates and prospects? In *The social life of nanotechnology*, ed. B. Her Hartorn and J.W. Mohr, 169–194. London: Routledge.

Powell, M., and D.L. Kleinmann. 2008. Building citizen capacities for participation in nanotechnology decision-making: The democratic virtues of the consensus Conference model. *Public Understanding of Science* 17(3): 329–348.

Public Debate Steering Board (PDSB). 2003. *GM nation? The findings of the public debate.* London: Department of Trade and Industry.

Radaelli, C.M. 2002. Democratising expertise? In *Participatory governance. Political and societal implications*, ed. J.R. Grote and B. Gbikpi, 197–212. Opladen: Leske and Budrich.

Ravetz, J. 1996. *Scientific knowledge and its social problems.* New Brunswick: Transaction.

Regonini, G. 2005. Paradossi della Democrazia Deliberativa. *Stato e Mercato* 60(1): 3–32.

Renn, O., and M.C. Roco. 2006. Nanotechnology and the need for risk governance. *Journal of Nanoparticle Research* 8(2): 153–191.

Roco, M.C. 2008. *Nanotechnology governance.* Presentation at the third international dialogue on responsible research and development of nanotechnology, Brussels, March 11–12.

Rowe, G., and L.J. Frewer. 2000. Public participation methods: A framework for evaluation. *Science, Technology and Human Values* 25(1): 3–29.

Stehr, N. (ed.). 2004. *The governance of knowledge*. New Brunswick: Transaction.

Weale, A. 2003. Science advice, democratic responsiveness and public policy. *Science and Public Policy* 28(6): 413–421.

Part III
Representations and Arrangements
of Responsibility

Chapter 8
Value Chain Responsibility
in Emerging Technologies

Colette Bos and Harro van Lente

8.1 Introduction

Corporate social responsibility (CSR) has gained visibility and importance for firms over the last decades. Increased pressures on firms to be more sustainable and responsible have resulted in many efforts inside and outside firms to consider societal responsibility and accountability. To support companies with this, several national governments and international organisations such as the ISO, have created multiple codes of conduct and voluntary guidelines for implementing and maintaining CSR within an organisation. Despite these efforts, the actual way CSR is filled in, differs from firm to firm. On top of that, because of the interrelatedness of value chains, companies are forced to consider the whole chain when implementing and exercising CSR. This so-called value chain responsibility (VCR) also shows diversity.

In general, CSR and VCR relate to established firms with known products and markets. This raises questions about the viability of CSR and VCR in the case of new technologies, where products, markets and even firms do not yet exist. In this chapter we investigate this question for the case of nanotechnology, a new technological field with many promising applications. A central characteristic is that substances have significantly different properties at the nano scale than they do at the micro or macro scale (Shelley 2006). These different properties and the small particle sizes may offer new functionalities, but have also caused reason for alarm because of possible health and environmental risks (de Jong et al. 2005). In addition, new technologies often come with questions about social, economical and political implications (Swierstra and Rip 2007). Both the health and environmental risks as well as broader societal and ethical questions about the applications of nanotechnology are currently a topic of debate.

C. Bos (✉) • H. van Lente
Copernicus Institute of Sustainable Development, Utrecht University,
Utrecht, The Netherlands
e-mail: C.Bos@uu.nl

S. Arnaldi et al. (eds.), *Responsibility in Nanotechnology Development*, The International
Library of Ethics, Law and Technology 13, DOI 10.1007/978-94-017-9103-8_8,
© Springer Science+Business Media Dordrecht 2014

Nanotechnology, as a new technology, thus raises new questions of responsibility and new implications for the CSR of companies. It raises the question how existing companies handle their CSR in the light of speculative technologies, but also how with new technologies new companies emerge. These new companies, in their turn, will have to form new value chain linkages, which will be speculative at the start.

This research thus intends to contribute to the current CSR and VCR literature which focuses on existing companies with existing technologies. We will describe how CSR is handled in more speculative and uncertain situations. We looked into this in three case studies of companies dealing with nanotechnology in their value chain in the Netherlands. All cases deal with nanotechnology, and differ in the stage of development of their company and value chain.

First, the current literature on CSR and VCR and the existing codes of conduct will be discussed. We then present and analyze the three cases. We conclude with a discussion on how our findings contribute to the current literature.

8.2   Theory

Most literature reviews mark the starting point of corporate social responsibility with the work of Howard R. Bowen from 1953 (Bowen 1953). Bowen noted that the largest companies at that time had the power to make decisions that influenced the lives of many citizens. He then argued that they should also take responsibility for their societal influence.

Despite the appreciation for Bowen's work, it took quite some time for firms to actually start implementing social responsibility practices. When in the 1980s and 1990s also the notion of sustainability started to gain more interest and this was connected to the responsibilities of companies, practices of CSR were established. From that time on, empirical studies have been conducted on how CSR was filled in (Carroll 1999).

Different studies have mapped the different views on CSR through history (Garriga and Melé 2004; Carroll 1999). An often used definition is: '(1) meeting objectives that produce long-term profits, (2) using business power in a responsible way, (3) integrating social demands and (4) contributing to a good society by doing what is ethically correct' (Garriga and Melé 2004).

This definition still leaves many blank spots for companies wanting to implement CSR practices. How should 'doing what is ethically correct' be filled in for instance, when situations can differ?

Pinkston and Carroll (1996) respond to this by arguing that CSR is elusive because it will differ with the values and beliefs of each point in time. To address these different values and beliefs, different stakeholders' perspectives can be taken into account (Snider et al. 2003). This corresponds well with the notion of considering the rest of the value chain when looking at CSR.

In general, the value chain can be described as the 'full range of activities which are required to bring a product or service from conception, through the different phases of production (involving a combination of physical transformation and the input of various producer services), delivery to final consumers, and final disposal after use' (Kaplinsky and Morris 2001, 4) When considering responsibility, it makes sense to adopt a broad view of the value chain that is not limited to companies which produce the product from raw material to the end-product, but also agents which are involved in other ways, such as through collaborations. In other words, it makes sense to consider the 'ecosystem' in which the value chain is embedded. Our case studies of value chains indeed start from an exploration of the whole 'ecosystem' in which the product is produced.

The need to consider the social responsibility of the entire value chain instead of just the CSR of the company itself, is something which has been addressed only in recent years. Phillips and Caldwell (2005) note this and argue that, adding to the notion of CSR, there is something such as value chain responsibility (VCR). They address the idea that the current large and international value chains have increased in size and power, but that the notion of a changing responsibility in these networks, is lagging. Increasingly, also the public demands that large companies take their responsibility for the entire value chain. One of the examples is Nike. Nike's manufacturing of clothing and running shoes was done by subcontracting companies in Asia, where bad labour circumstances were found – e.g. child labour and slave-like conditions. Nike argued that they could not change these circumstances since they did not own these companies; they were merely suppliers to Nike. The unwillingness of Nike to change this, however, led to large public protests among the USA and to some large university sporting teams to boycott the use of Nike sportswear. Nike eventually gave in and made agreements with the suppliers to improve the labour situations.

The interesting notion here is that Nike did not have any legal responsibility or formal liability in this case. However, since they are seen as a powerful player the public opinion was that they were responsible anyhow. A powerful player is able to make changes for the better, according to the popular argument. Nike eventually gave in to this, not only because the boycotts decreased their profits, but mainly because of the reputation damage that was being done.

The work of Phillips and Caldwell (2005) mainly addresses examples of large and well-known companies such as Nike, McDonalds, Gap, and Starbucks. They consequently recommend that more work is needed to study the VCR of smaller firms or firms with a less public profile. While the responsibility issues might be less visible in firms with little public profile, they do exist and are interesting to look into.

Currently, most empirical studies on CSR investigate the link between competitive advantage and CSR (McWilliams and Siegel 2001; Porter and Kramer 2006; Sen and Bhattacharya 2001). Other research does follow up this research with more strategic implications, including new technological innovation. McWilliams et al. (2006) here show with the example of British textile industry in the nineteenth century that new technologies can change the social responsibility perspective. The British textile industry was namely only able to abandon child labour practices after new machines were introduced. In general, however, the notion that CSR and VCR could change due to new technological developments is lacking in the current literature. Some articles do address the topic of how to innovate – including creating new value chain linkages – while being socially responsible (McWilliams et al. 2006; Luo and Bhattacharya 2006). Still, the current literature on CSR and VCR mainly describes

Fig. 8.1 Literature gap in
VCR and CSR literature

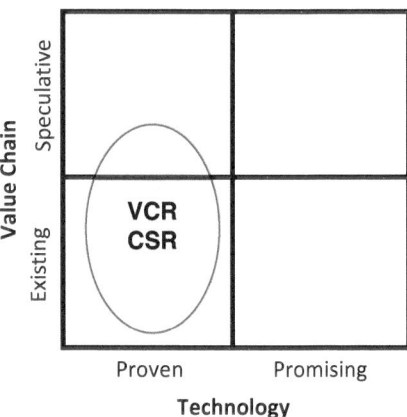

practices of firms with proven technologies in existing value chains, where sometimes these practices might be extended to new and more speculative value chains.
In this chapter we will make a further step and investigate the issue of social
responsibility in emerging technologies and speculative value chains. While so-
called studies of ethical, legal and social aspects (ELSA) of promising technologies
tend to focus on policy purposes, we address the responsibilities of firms and seek to
extend the current CSR and VCR literature to new technological development and
the value chains that emerge with this (Fig. 8.1).

In doing so, our research also follows up on the recommendations of Phillips of
Caldwell (2005) to look at also smaller firms. This combines well, since new technological developments often cause new and small technological start-ups, where
the value chain is still highly speculative.

8.3 Formalised Forms of CSR

Responsibility is a difficult concept for companies to deal with. Individual responsibility
is something which most people can grasp and decide upon, based on their individual
moral beliefs and values. But it is difficult to extend this individual concept of responsibility to the responsibility of a company. Two other aspects also come into play when
looking at the company's responsibility: it has to be communicable among the whole
company and it has to relate to practicalities of everyday practices.

For these reasons, several formalised forms of CSR have been developed during
the last decades. Also governments and supranational bodies such as the EU have
been stimulating the development and implementation of these guidelines. A guideline does not only make responsibility communicable within the company, but also
makes clear to the outside world how the company is handling its CSR. This means
that the CSR can be evaluated.

Since this research focuses on companies and value chains in the Netherlands,
some Dutch national initiatives for CSR are described here. Furthermore, different
international guidelines are added.

The Dutch Corporate Governance Code (CGC), which used to contain only internal good governance practices, has now been extended with corporate social governance guidelines. Every company that is stock listed has to comply with the CGC, which means they yearly have to report how they implemented the principles from the CGC. This code is non-binding: the companies do not have to comply with all the principles, but if they do not, they have to justify this (Dutch Corporate Governance Code 2008).

Another initiative is the Transparency Benchmark, which is initiated by the Ministry of Economic Affairs, Agriculture and Innovation. This is not a CSR guideline in itself, but it obliges companies to be transparent in the CSR activities and reporting. The 500 largest companies of the Netherlands are included in the research group. Smaller companies can also participate on a voluntary basis (Transparantie Benchmark 2013a).

Furthermore, the Dutch Social and Economic Council produced some advice for the Dutch government on sustainable globalisation, which also addressed the value chain responsibility of companies (SER 2008). The Dutch government followed this advice and made a recommendation for CSR for companies which are facilitated by the government. This advice gave some general guidelines for CSR and also included the notion that the value chain should be seen as a primary responsibility (van Heemskerk 2008).

Internationally, several CSR guidelines have been developed. Recently the ISO (international Organisation for Standardisation) has made the ISO 26000 code, '*Guidance on social responsibility*' (ISO 2010) which gives information for organisations to understand social responsibility and practical guidelines for implementing it. ISO 26000 is a voluntary guideline for companies and is also based on self-assessment, so a public evaluation of it may be difficult.

Since the responsibility of nanotechnology in particular could relate to the possible health and environmental concerns, guidelines for the use of certain substances are relevant as well. Regarding possible harmful materials, the REACH list is the most used standard. It deals with the Registration, Evaluation, Authorisation and Restriction of Chemical substances (EC 2013). The REACH standards are made and controlled by the European Chemical Agency (ECHA). Also nanosubstances fall under the scope of REACH, because any potentially harmful substance needs to be submitted to REACH regulation, irrespective of its size. However, the following is noted in the Frequently Asked Questions about REACH:

> The evolving science of nanotechnology may necessitate further requirements in the future to reflect the particular properties of nanoparticles (ECHA 2013).

This suggests that the ECHA is considering whether the current REACH legislation is broad enough to include nanosubstances or that a new legislation needs to be developed. Companies will thus have to decide for themselves what they consider a responsible way to deal with these nanoparticles.

Apart from these guidelines, companies can also have guidelines that they have developed themselves or relate to more informal CSR practices. In small companies, the need for guidelines will be less, since the CSR of the company will then be closely intertwined with the individual responsibility of the employees.

8.4   Case Studies

To study how social responsibility appears in companies which deal with nanotechnology, three cases were selected. The first case refers to a large company, which already has an established value chain, even though the product they are developing is new. The second case has its value chain in place, in principle, but there is room for company growth and value chain extension. The last case is a small starting company, where value chain linkages are still very unclear.

First, interviews were conducted within the case companies. From these interviews, the value chain links were revealed and mapped. Secondly, we interviewed many other agents in the value chain to triangulate our findings on social responsibility.

All interviews were recorded and were consequently transcribed. From these transcriptions, descriptive case outlines were made. These described how the companies filled in CSR and VCR with arguments, routines and practices.

8.4.1   Social Responsibility in Different Stages of Development

The first case company is Philips, a large organization, which used to focus on consumer electronics, but has shifted the emphasis to health and wellbeing technology. Now healthcare products represent about 40 % of Philips's business revenues. Philips is a large and international company, with total assets of over 30 billion Euros and 119,000 employees in more than 60 countries. One of their recently established business ventures, hand-held diagnostics, seeks to develop a handheld diagnostic device based on the use of nanoparticles: the Magnotech concept.

The technology is based on magnetic nanoparticles, which are coated with the appropriate ligand molecules for the target protein molecules that are to be measured. A small sample of blood or saliva can be inserted in a cartridge which can in turn be inserted in the device. In the device the magnetic nanoparticles bind to the target protein molecules in the sample blood or saliva. By turning magnets in the device on and off, a fast and accurate separation between the bound and unbound particles can take place (Philips.nl 2013a).

Even though the product has not yet reached the market stage, it can profit from being within the large network of Philips. In the supplier field, links with the suppliers of the nanoparticles have been made. Philips has chosen to not produce the nanoparticles themselves, since this is not their core capability. When the product reaches its marketable phase, it can also profit from Philips' existing value chains with its market and sales channels. For this product, Philips already has ties with many hospitals and medical professionals, because of their earlier healthcare products. Figure 8.2 shows the value chain of Philips Magnotech.

Figure 8.2 also illustrates that Philips is involved in many collaborations, both bilaterally (e.g. with suppliers and partners) and within public-private partnerships such as CTMM (Centre for Molecular Medicine). Here they also have ties with multiple hospitals (in this research UMCU and AMC participated).

Fig. 8.2 Value chain
of Philips Magnotech

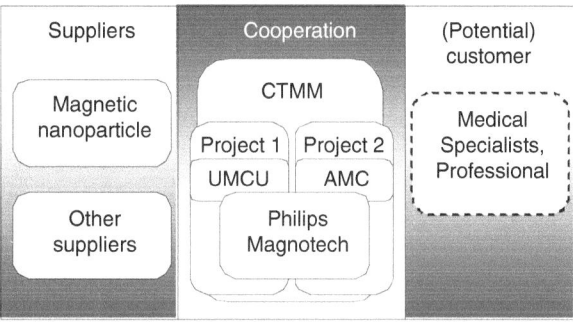

CSR within Philips is addressed with a formalised procedure. They have an elaborate sustainability program, which is renewed every 5 years, called EcoVision (Philips 2013b, c). Ecovision includes environmental and societal aspects. Philips is also incorporated in the Transparency Benchmark, in which it has scored in the top-20 since 2005. Philips was ranked in the second place in 2010 and 2011 and in the third place in 2012 (Transparantie Benchmark 2013b).

Furthermore, Philips has formalised procedures for organising responsibility among the value chain. Every supplier of Philips has to agree to the Supplier Sustainability Declaration that Philips has. In this declaration similar guidelines as in the EcoVision program are addressed (Philips 2012). Philips also audits the suppliers to see whether they actually live up to these standards.

The interview brings forward how the EcoVision program sets preconditions that assure that social responsibility is handled correctly. This means that the individual employees do not have to be concerned with CSR much:

> To say that it would have an influence on day-to-day activity, would go a bit too far.

Furthermore, the power of Philips as a large company is apparent when referring to their suppliers and VCR. Philips has the power to demand social responsibility from their suppliers:

> With partners social responsibility needs to be negotiated (...) But this is very different with suppliers: those you can just impose your policy on.

Philips's strict control of their own social responsibility and that of their value chain is intertwined with strategic considerations about reputation. Philips is a large and well-known company and does not want its name to be damaged, in the way Nike's was in the earlier example. Furthermore, being actively involved in social responsibility also helps to positively affect Philips' reputation. Philips 'advertises' with their sustainability and responsibility goals on their website.

For Philips, being involved in a nanotechnology value chain now does not change their CSR and VCR perspective. Concerning health and environmental risks of nanotechnology, they rely on REACH regulations. Even though some new value chain linkages had to be made for this new technology – i.e. with the nanoparticles suppliers – Philips has all the control over its own and its value chain responsibility.

In our second case, TSST, we studied a much less developed value chain. TSST is a small company, which began as a university spin-off in 1998 (TSST 2013). The University

Fig. 8.3 Value chain of TSST

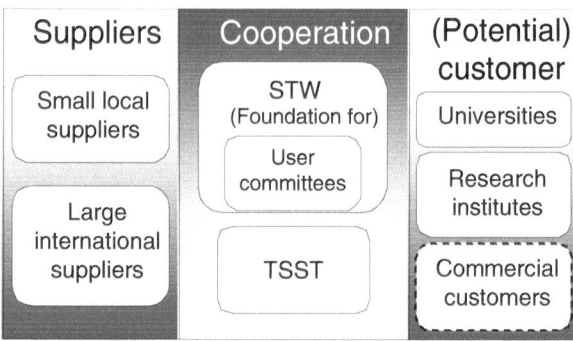

of Twente needed specific deposition equipment for their nanotechnology research and since they could not find a suitable supplier, the science department decided to have some employees develop and build the machine themselves. The final machine turned out to be of such good quality that universities and research institutes all around the world requested to also have one built. Thus a spin-off was created to develop these Pulsed Laser Deposition Systems. These machines are currently one of the main research instruments for conducting nanotechnology research.

The machines which TSST builds are specifically built to customer preferences and take approximately 6 months to make. The supplier side of the value chain of TSST consists of roughly two groups: suppliers for ready-made catalogue parts, these are usually large and international companies and specifically designed machine parts; these are local and smaller companies. In their start-up phase TSST was within a project of STW ('Stichting Technische Wetenschappen', in English: Foundation for Technical Science). STW links university research which has commercial potential with so-called 'user-committees'. These committees comprise of potential interesting consumers for the technology. The customers of TSST are all universities or research institutes from all over the world. They, however, do not have regular customers because it is rare that a consumer needs a second machine. For the first time in its existence TSST is now considering requests from commercial companies instead of research institutes or universities. It is still uncertain whether TSST will start supplying to these commercial companies (Fig. 8.3).

TSST does not have formal regulations regarding social responsibility. Moreover, they indicate that social responsibility is not a very salient or complicated item. The reason for this is that TSST operates in very controlled and trusted value chains:

> We operate in the scientific world. Everybody knows everybody. (…) So there is a lot of social control.

That they trust their customers also stems from the fact that the machines are so very specifically built for research purposes:

> You cannot do anything else with these machines than the research that they are specifically built for.

The implication here is that if nothing else can be done with it, also nothing 'bad' can be done with it. The trusting relationship with the other value chain actors is thus based on both the social relations and the specificity of the machine.

Fig. 8.4 Value chain of MyLife Technologies

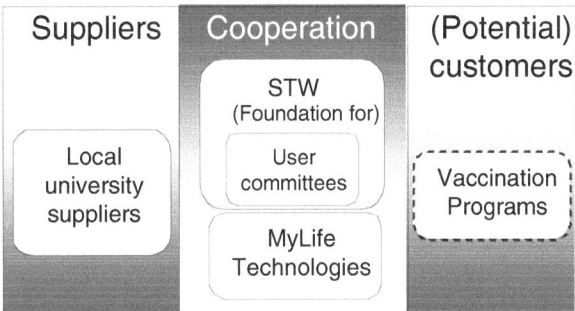

For TSST, their responsibility is to deliver sound machines to customers who they trust. This makes their value chain very stable. They do not think their responsibility extends to their customers where the 'actual' nanotechnological work is conducted. This could change however if they decided to start supplying for commercial customers.

The third company we studied is MyLife Technologies BV. At the time of the empirical investigation (Spring/Summer 2011) MyLife Technologies was a company in the making; it had not yet been registered at the Dutch Chamber of Commerce, but was still a university research group with the wish to commercialize its research. In February 2012, MyLife Technologies succeeded in becoming a registered company. The empirical results presented here concern the time when MyLife Technologies was not yet a formal company.

MyLife Technologies has developed, and is still developing, microneedle arrays (MNA). These are needles with micrometer dimensions, over 50 times shorter than a classical syringe. Since they only penetrate the outer skin but do not reach the sensitive nerves which lie below, this makes the application painless. The small volume of the needle is compensated by placing multiple microneedles in an array. This array is then integrated in a patch (MyLifeTechnologies.nl 2013).

The patches are easier to handle than classical syringes. This and the painless application could make the patches good candidates for vaccination of children. Another possible application is the combination of the patches with Lab-On-A-Chip technologies to create easy ways to self-test certain levels in the blood. Furthermore, the technology is developed further to see if there are possibilities for 'intelligent patches' which could, for example, contain more also newly developed 'nanomedicine'. This would make the patches a totally new way of drug delivery combined with specifically designed new drugs for this way of delivery, instead of just creating a new way of delivering existing drugs.

This is still a very young company and not many value chain linkages have been made. While ideas about applications and products circulate, the chain is still very speculative. Some linkages have been made in research collaborations and also MyLife Technologies is part of a project from STW. Through this project a vaccination program has showed interest, but no concrete steps have been made yet. Not only is the value chain speculative, but also the final form of the technology will depend a lot on which customers MyLife Technologies is able to find (Fig. 8.4).

MyLife Technologies has no formal forms of CSR, which is not surprising since the company does not even formally exist yet. The speculative nature of the future of MyLife Technologies renders its social responsibility speculative as well:

> Corporate social responsibility will come when we really form a company.

However, notions of a certain responsibility do play a role.

> The responsibility for a new technology lies with the people who initiate the idea. But if you believe that this is a technology that could help people, I think you also have a certain responsibility when you don't do it.

For MyLife Technologies it matters a lot that they are in the field of nanotechnology. Their chances of commercialising their product could be seriously limited if the public opinion on nanotechnology were negative. This is also the reason why they want to try to pursue the option of using the microneedle arrays for vaccination:

> Vaccination is something established. It is initiated from the government and it is socially accepted. This would be a good way to let people come into contact with our MNA patches.

MyLife Technologies clearly deals with a lot of speculation. Both their technology and value chain have not found a stable form yet. Furthermore, the final form of the value chain and technology are also interdependent. They are, however, very responsive to the fact that they are dealing with nanotechnology and are already considering some implications for responsibility issues beforehand. This could prove useful in developing further CSR measures when the technology and value chain stabilise.

8.5   Responsible Speculation

This chapter intends to investigate how to deal with responsibilities in the case of new technologies. In the last decade, the tradition of corporate social responsibility (CSR) of companies has been elaborated. In recent years this perspective on CSR has been extended with considering the responsibility of the value chain, which resulted in the notion of value chain responsibility (VCR).

We discussed how the CSR and VCR literature mainly addresses existing technologies and stable value chains. While these approaches sometimes are extended to possible new value chains, they largely ignore the issue of how firms (should) handle their social responsibility in the light of emerging technologies and the speculative value chains that accompany them.

In our study we explored these new areas of investigation. Figure 8.5 shows how the cases in this study can be positioned amongst the axes of existing versus emerging technology and the stable versus speculative value chain. We think that our empirical cases show that the expansion of the current CSR and VCR literature is sensible.

Philips is currently engaged in a new technology with their Magnotech project, but it is clear that they are also still set in stable value chains with existing technologies. Also for the new technology, the value chain is clear and stable. This stability – based

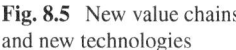

Fig. 8.5 New value chains and new technologies

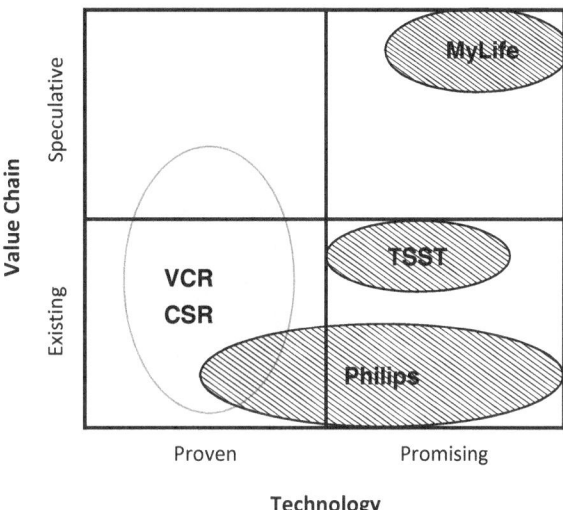

on previous products – resonates in their approach to CSR. Both CSR and VCR are filled in formal and controlled fashions. Because of the stability of their social responsibility approach, not much changes when they deal with a nanotechnology product. Philips relies on their CSR framework and on the REACH regulations.

Even though TSST is young company, their value chain is relatively stable. They do not have the power to have the controlled and very stable value chain of Philips, but their stability comes from the trusting relations with the actors. As a university spin-off, TSST remained within the social network of the academic world, where they know and trust their customers. They produce a new technology, but since the purpose of this technology is clear and limited, this also does not bring about much speculation. The newness of the technology does not have much impact on responsibility issues here, since the value chain is quite stable.

MyLife Technologies is clearly in the top right corner of the spectrum in Fig. 8.5. Both the technology is very new and the value chain is still highly speculative. Furthermore, the final form of the technology and the value chain depend on each other. Because of this uncertainty, also the responsibility issues are vague and unspecified. On the one hand, MyLife Technologies sees itself as unable to consider its social responsibility until the value chain becomes more stable and until what this means for their responsibility can be evaluated. On the other hand, dealing with an emerging technology such as nanotechnology, means that MyLife Technologies carefully considers public opinion and hopes to influence it. Their wish to link their technology with the vaccination program can be seen as an attempt to both stabilise the technology and the value chain speculation.

So it seems that when companies deal with both new technologies and new value chains, this can lead to changes in their view on social responsibility. When dealing with only a new technology, but in a stable value chain, the company can still rely on this stability of the value chain and not much change in social responsibility is deemed necessary.

This leads to interesting questions for further research into value chain responsibility. Within highly speculative value chains, also the considerations about the responsibility within that chain can only remain speculative. But when is the moment that the chain stabilises enough to consider the company's and chain's responsibility? And how does this relate to the responsibility that other value chain actors take and the mutual expectations in this value chain?

Something else to consider is that while CSR and VCR approaches can be enriched by adding notions of emerging technologies and speculative value chains, these approaches, in their turn, can enrich the debate on emerging technologies. Since nanotechnology is expected to be a disruptive technology – and because of the possible health and environmental risks – increasingly more ethical and social science research is directed to the 'responsible embedding' of nanotechnological applications in society. This research focuses mainly on either theoretical ethical considerations or responsible practices of nanotechnology researchers, since in most cases nanotechnology research has not yet reached commercial applications yet (See e.g. McGinn 2008; Roache 2008).

These studies address questions like whether we, as a society, would or should want new technologies, which could potentially be harmful for humans and the environment. Also unintended consequences could result from releasing artificially produced nanoparticles into the air, soil or water systems. While these are legitimate questions and considerations, they do not give any clear instructions for firms wanting to handle these technologies responsibly.

The cases in this research however, show that when applications do move from university research to commercial applications in companies, a more practical form of the responsible handling of nanotechnology is needed. CSR practices and guidelines could perhaps function as examples for the ethical work to become more 'action-guided' and move out of the theoretical realm. This could consequently lead to adding these ethical considerations to CSR guidelines, to embed them better in ethical considerations.

The empirical results from this study thus show that the CSR and VCR literature could be enriched by including cases of new technologies and speculative value chains. On the other hand, it seems that studies of ethical, legal and social aspects (ELSA) of new (nano)technologies could benefit from commercially embedded approaches of CSR initiatives. Both strands of thinking could prove valuable for firms currently dealing with the demand to be both socially responsible and innovative.

References

Bowen, H.R. 1953. *Social responsibilities of the businessman*. New York: Harper & Row.

Carroll, A.B. 1999. Corporate social responsibility: Evolution of a definitional construct. *Business and Society* 38(3): 268–295.

de Jong, W.H., B. Roszek, and R.E. Geertsma. 2005. *Nanotechnology in medical applications: possible risks for human health*. RIVM report. 265001002/2005. Bilthoven.

Dutch Corporate Governance Code. 2008. *Principles of good corporate governance and best practice provisions*. The Hague: Corporate Governance Code Monitoring Committee.

European Chemicals Agency (ECHA). 2013. *Frequently asked questions about REACH.* Q&A unique ID: 0010. Accessed 18 Nov 2013.

European Commission. 2013. *European Commission website on REACH.* http://ec.europa.eu/enterprise/sectors/chemicals/reach/index_en.htm. Accessed 18 Nov 2013.

Garriga, E., and D. Melé. 2004. Corporate social responsibility theories: Mapping the territory. *Journal of Business Ethics* 53(1–2): 51–71.

ISO. 2010. Guidance on Social Responsibility - ISO 26000:2010(E). Geneva.

Kaplinsky, R., and M. Morris. 2001. *A handbook for value chain research*, 113. Ottawa: IDRC.

Luo, X., and C.B. Bhattacharya. 2006. Corporate social responsibility, customer satisfaction, and market. *Journal of Marketing* 70(4): 1–18.

McGinn, R. 2008. Ethics and nanotechnology: Views of nanotechnology researchers. *NanoEthics* 2(2): 101–131.

McWilliams, A., and D. Siegel. 2001. Corporate social responsibility: A theory of the firm perspective. *The Academy of Management Review* 26(1): 117–127.

McWilliams, A., D.S. Siegel, and P.M. Wright. 2006. Corporate social responsibility: Strategic implications. *Journal of Management Studies* 43(1): 2–17.

Mylifetechnologies.nl. 2013. http://www.mylifetechnologies.nl/. Accessed 18 Nov 2013.

Philips. 2012. *Royal Philips electronics, supplier sustainability declaration.* CSO-BPOI-2012-005. Eindhoven. http://www.philips.com/shared/assets/company_profile/downloads/EICC-Philips-Supplier-Sustainability-Declaration.pdf

Philips.com. 2013a. *Magnotech.* http://www.business-sites.philips.com/magnotech/technology/index.page. Accessed 18 Nov 2013.

Philips.com. 2013b. *EcoVision 4.* http://www.annualreport2009.philips.com/pages/our_group_performance/sustainability/ecovision4.asp. Accessed 18 Nov 2013.

Philips.com. 2013c. *EcoVision 5.* http://www.annualreport2009.philips.com/pages/our_group_performance/sustainability/ecovision5.asp. Accessed 18 Nov 2013.

Phillips, R., and C.B. Caldwell. 2005. Value chain responsibility: A farewell to arm's length. *Business and Society Review* 110(4): 345–370.

Pinkston, T., and A. Carroll. 1996. A retrospective examination of CSR orientations: Have they changed? *Journal of Business Ethics* 15(2): 199–207.

Porter, M.E., and M.R. Kramer. 2006. Strategy and society. The link between competitive advantage and corporate social responsibility. *Harvard Business Review* 84(12): 78–92.

Roache, R. 2008. Ethics, speculation, and values. *NanoEthics* 2(3): 317–327.

Sen, S., and C.B. Bhattacharya. 2001. Does doing good always lead to doing better? Consumer reactions to corporate social responsibility. *Journal of Marketing Research* 38(2): 225–244.

Shelley, T. 2006. *Nanotechnology: New promises, new dangers.* London: Zed Books.

Snider, J., R.P. Hill, and D. Martin. 2003. Corporate social responsibility in the 21st century: A view from the world's most successful firms. *Journal of Business Ethics* 48: 175–187.

Social and Economic Council of The Netherlands (SER). 2008. *Advies Duurzame globalisering: een wereld te winnen.* The Hague: SER.

Swierstra, T., and A. Rip. 2007. Nano-ethics as NEST-ethics: Patterns of moral argumentation about new and emerging science and technology. *NanoEthics* 1(1): 3–20.

Transparantie Benchmark. 2013a. *(In English: Transparency Benchmark) about the Transparantie Benchmark.* http://www.transparantiebenchmark.nl/Over%20Transparantiebenchmark. Accessed 18 Nov 2013.

Transparantie Benchmark. 2013b. *(In English: Transparency Benchmark) on the CSR transparency. List of companies.* http://www.transparantiebenchmark.nl/bedrijven. Accessed 18 Nov 2013.

Twente Solid State Technology. 2013. *TSST.nl.* http://www.tsst.nl. Accessed 18 Nov 2013.

van Heemskerk, F 2008. *Brief van de staatssecretaris van Economische Zaken aan de Voorzitter van de Tweede Kamer der Staten-Generaal* [Letter from the State Secretary of Economic Affairs to the Chairman of the Parliament]. Den Haag, 22 Dec 2008.

Chapter 9
On Being Responsible: Multiplicity in Responsible Development

Sarah R. Davies, Cecilie Glerup, and Maja Horst

9.1 Introduction

We are, in this essay, concerned with a single central question. What does responsibility look like? Like motherhood and apple pie, it is very difficult to argue _against_ responsibility. No-one wants to portray themselves as irresponsible, and most people – indeed, most organisations and businesses, too – are very happy to align themselves with responsibility as a central principle for their conduct. But what does this actually mean, in practice, in different contexts? How is responsibility performed and understood? In considering this we want to describe a couple of instances in which 'being responsible' is being worked out, and to reflect on their differences. Responsibility, we will argue, is fundamentally multiple and contingent. While many people, organisations, and actors may lay claim to being responsible, using rather homogeneous language, its performance is always shaped by the dynamics and pressures which act upon a particular site at a particular moment.

In doing this we will focus on one area in which responsibility is currently a key emphasis: that of research and development in nanoscience and nanotechnology. In discussing these questions we are thus building upon a small body of work which has started to unpick the current drive towards responsibility in nanotechnology's development. At the most basic level, such work has *documented* this drive, showing

S.R. Davies (✉) • M. Horst
Department of Media, Cognition and Communication,
University of Copenhagen, Copenhagen, Denmark
e-mail: dxq327@hum.ku.dk

C. Glerup
Department of Organization, Copenhagen Business School,
University of Copenhagen, Copenhagen, Denmark

S. Arnaldi et al. (eds.), *Responsibility in Nanotechnology Development*, The International
Library of Ethics, Law and Technology 13, DOI 10.1007/978-94-017-9103-8_9,
© Springer Science+Business Media Dordrecht 2014

how nanotechnology – the interdisciplinary research field composed of 'science, engineering, and technology conducted at the nanoscale' (NNI n.d.) – has become marked by the language of 'responsible development'. While the history of nano-technology is a relatively brief one (if we mark its inception by major research funding programmes, such as the US's National Nanotechnology Initiative, signed into being in 2000, or the production of policy statements, such the 2004 European Commission Communication *Towards a European Strategy for Nanotechnology*), discourse around the technology has rapidly taken on a number of distinctive features. Nanotechnology has been framed as an *opportunity*: for the development of new modes of governance which move from 'command and control' to more distributed methods of policy coordination; for the integration of social science and humanistic research into natural science research programmes; for public participation and engagement to shape policy and research funding (Jones 2008; Macnaghten et al. 2005; Roco et al. 2011; Stilgoe et al. 2013). Most of all, nanotechnology is por-trayed as an opportunity to *get things right* – to avoid the mistakes made in attempts to commercialise genetically modified crops and thus to enable economic growth through technological innovation (Randles 2008, see also European Commission 2004). Responsible development, it is argued, is the way to achieve this (Krupp and Holliday 2005).

Beyond this characterisation of the 'governance landscape' of nanotechnology – one pitmarked by the craters of failed voluntary reporting schemes, and washed about by the waters of public participation (Kearnes and Rip 2009) – a number of studies have sought to explore how responsibility is being discussed, operationalised, and mobilised in practice. Kjølberg and Strand (2011), for instance, took the European Commission recommendation on a code of conduct for responsible nanosciences and nanotechnologies research (or EC-CoC, European Commission 2008) as their starting point, conducting interviews with nano-researchers in order to discuss the text of the code. Significantly – given that the EC-CoC has been a major European effort towards enabling responsible nano research – they found that none of the researchers they spoke to had previously heard of it; they also charted a significant degree of diversity in the ways in which researchers responded to its seven 'principles' (of meaning, sustainability, precaution, inclusiveness, excellence, innovation, and accountability). This diversity extended to the ways in which their interviewees understood what it meant to be responsible: while, as they write, '[a]ll the researchers described them-selves as responsible, and shared the view that researchers in general are responsible' (shades, again, of motherhood and apple pie), when probed, this was expressed in 'quite different views about what this means in terms of what scientists are responsible for, and who they are responsible towards' (Kjølberg and Strand 2011, 103).

Such views included the sense that one's responsibility is, first and foremost, to the science, which should be both excellent and free of societal constraints; an emphasis on the need for risk research; and a 'social democrat' perspective in which regulatory authorities should – with the aid of public communication and consultation – define the direction nanoscience and technology takes.

McCarthy and Kelty (2010), in a longer term ethnographic study of the Center for Biological and Environmental Nanotechnology (CBEN) at Rice University, are similarly interested in how responsibility is being articulated – in how it is 'made 'do-able" (McCarthy and Kelty 2010, 406) for scientists working in nanotechnology. They treat responsible development as, essentially, a practical and very concrete challenge for those studying nanoscience, tracing CBEN's history from its lead researcher's desire for a transmission electron microscope (TEM) and subsequent Center funding bid to the inclusion of 'implications' research into this Center and the mobilisation of 'responsibility' as a means of both securing research funding and ensuring public support for a burgeoning industry. Ultimately, McCarthy and Kelty write,

> the fact that there is a widely felt demand for science to become more responsible does not prescribe what that responsibility will eventually look like, or the effects it will have. [...] Responsibility is thus approached as something emergent and historically contingent, which is given stability through the practical work of creating new institutes, centers, and councils, reorganizing scientific work, and re-thinking the relationships between disciplines. (McCarthy and Kelty 2010, 426–427)

Responsibility, in their case study, is something to be constructed, manipulated and pushed around: as their ethnographic engagement closes they observe the location of responsibility being shifted such that scientists can continue their science whilst an external entity – the International Council on Nanotechnology, ICON – acts as a 'filter' by which societal questions may be purified into scientific ones. Significantly, notions of public acceptance (or not) are an important driver within these moves, with those involved in the development of CBEN and ICON construing risks both *of* and *to* nanoscience: the central dynamic is that, if scientific work on the implications of nanoscience for human health and the environment is not pursued, 'public backlash will grow and itself become a risk to nanotechnology' (McCarthy and Kelty 2010, 421). Similarly, Shelley-Egan (2010), in a study exploring divisions of moral labour within nanotechnology's development, cites 'credibility pressures' as a central force acting upon industrial actors in managing calls for responsibility. Here, again, the 'risk' of public concern shapes the operationalisation of responsible development (cf. Shelley-Egan and Davies 2013).

A number of studies have, then, begun to indicate the contingency of responsibility. It is this theme of multiplicity that we wish to develop in the rest of this essay, exploring the diversity of articulations of responsibility in a number of sites. We start by discussing more thoroughly responsibility in the academic literature of sociology of science, organisation studies, and higher education, showing that there is multiplicity even in a domain often considered more stable than 'real world' practice. These framings are then related to the ways in which notions of responsible development are understood, and acted upon, in two different US sites: an academic research centre, and the nanotechnology private sector. We do this by telling the story of a particular project of the NSF-funded Arizona State University-based Center for Nanotechnology in Society (CNS-ASU), in which its notions of responsibility in nanotechnology came into contact with those of the private sector. We argue that the multiplicity of responsible development is brought into particular focus by such clashes of taken-for-granted meanings and values.

9.2 Responsibility in the Literature

How, then, is responsibility conceptualised in the academic literature? In this section we draw on an extensive keyword based search, conducted in 2011, of journals focusing on higher education, research ethics and the sociology of science, which sought to elicit articles discussing social responsibility in scientific practice. From this search approximately 250 articles were identified, and a discourse analysis conducted on their content and arguments (see Glerup and Horst 2014). From this analysis we want to draw out a number of key dynamics, around which the literature appears to be structured: firstly, that the concept of responsibility of science appears largely to be connected to the notion of *obligations* (such as, for instance, the need to contribute to economic growth, avoid research misconduct, engage the public, or provide solutions to social problems); secondly, that differing perspectives on responsibility are grounded on differing diagnoses of the '*problems*' that warrant responsibility; and, thirdly, that the central relationship within which these problems are diagnosed is that between *science and society*.

The bulk of the articles identified in the literature search thus present normative arguments for 'more' or 'a different form of' responsibility in science by describing ways in which something is 'wrong' with science or with the ways in which it relates to society. Given this central perspective, the literature can be divided according to two key, opposing perceptions of the problem that triggers the need for responsibility within the science-society relationship. The first insists that problems stem from science having become *too entangled* with society, such that a clearer boundary between science and society is necessary. The second suggests the opposite: that science is *too cut off* from wider society, and needs to be better connected to it.

The first position might be summarised as a 'demarcation discourse', and enrols a classic, modernistic view of science. In this discourse, the problem with current science and scientific practice is that interference from society is threatening freedom of research as well as the ideal of disinterestedness. Implicitly – or, at times, explicitly – this discourse refers to Mertonian norms (see Merton 1973): the central concern is to keep science from being polluted by non-scientific interests, concerns, or influences. Only by keeping science pure, the argument goes, is it possible to produce unbiased truth – an objective which is considered both the central obligation of science and the way in which science can best serve society. One example of this discourse can be found in Deichmann (2005), who argues that new tendencies to demand 'responsibility', or make science work for 'the greater good', are downright dangerous. According to Deichmann, notions such as the greater good are an unstable human invention, and therefore all too easy to manipulate. Science should focus on facts, and leave it to politicians to decide whether and how scientific knowledge can be used (cf. Wolpert 2005). In this discourse, then, the heart of responsible science involves enforcing internal discipline and a rigorous scientific method. Discipline and method serve as a protective shield against social influences, and it is therefore the duty of scientists to practice them in the most rigorous way.

Within this demarcation discourse responsibility relies on cutting off, as far as is possible, all ties of influence and connection between society and scientific practice. In contrast, a large number of the articles which treat responsibility diagnose the problem in exactly the opposite way: they call for a more pronounced connection between science and society. Within these articles three distinct perspectives emerge, depending on how they call for this stronger science-society relationship to be structured. We have summarised these perspectives according to their foci on notions of usability, reflexivity, or democracy. Each of these discourses of responsibility are discussed below.

In the 'usability discourse' the call for responsibility is predicated on the notion that, currently, science does not deliver sufficiently useful results to society. Within these calls for useful, or usable, research a number of different definitions of usefulness can be detected. In Cross and Price (1999), for instance, the problem of lack of usefulness is connected to a problem of communication between science and the population. Their articulation of responsibility as usability thus draws on what could be called a traditional model of public understanding of science (Durant et al. 1989; Michael 2002), in which publics need to understand science in order to be empowered to make use of it. As such, usefulness can be understood as relating to concepts of science as a public good: science's fundamental purpose is to enable better lives, and citizens and other social actors should be able to use it for this purpose. Notions of usability, however, are also connected to ideas of the knowledge economy and the need for science to be a driving force in the creation of globally competitive societies, in which economic growth is based on university-industry collaboration. Here, usefulness is closely linked to innovation; that is, to ideas, concepts and technologies that can be transformed into products on the market (Barbosa and Faria 2011; Corolleur et al. 2004). Usefulness is thus something that contributes to economic growth and which provides competitive advantages for countries, regions, or organizations.

Whether usefulness is understood in relation to public understanding of science or to infrastructures for innovation, this literature is united in arguing that there is a lack of sufficient relations between science and the rest of society. Both strands of work which present this usability discourse focus their solution on some form of political management, distributing responsibility such that it incorporates actors other than science and scientists. Some texts argue that the media are in part responsible for enabling usability, for instance, as they have a duty to report and discuss science and technology and thus to help create a public sphere in which these issues can be debated (Kowal 1980). Others focus on the creation of supportive frameworks such as the triple helix (Leydesdorff and Etzkowitz 1998). What unifies the texts in this discourse is thus their focus on usefulness as a prerequisite for responsible science, and the fact that they argue for the involvement of external actors such as legislators, governments, citizens and experts in bringing about – and applying – this useful and usable scientific knowledge.

The usability discourse focuses on potentially positive outcomes of science: its emphasis is on the benefits of technoscience, whether directly, as citizens understand, take it up, and make use of it, or indirectly, as it is applied by industry and brings about economic development. In contrast, the two final discourses of responsibility we want to discuss (of 'reflexivity' and 'democracy') are structured around a focus on potential negative outcomes of the development of science and technology. Both deal with the need to engage more systematically with intended as well as unintended effects of science on society; they are distinguished, however, by the extent to which they favour external involvement and regulation of science. One set of arguments points to *scientists* as the key actors who need to reflect on the possible consequences of science. The other suggests that these issues cannot be left to scientists alone, and that science needs to come under *democratic control*.

We have summarised arguments that focus on internal regulation as a 'reflexivity discourse'. The thrust of this literature is that science is expected to be able to regulate itself (much as in the demarcation discourse discussed above). However, the reflexivity discourse argues that it is important, as part of the process of doing science, to consider the consequences of science and technology and to take responsibility for the fact that some of the problems of modern society have been caused by technoscientific development. On this basis, science has an obligation to reflect on its own effects, and to regulate itself so as to account for its social impacts (Forge 2000). Work which makes use of this discourse might point to the importance of including training on social responsibility when teaching university students (Gilmer and DuBois 2002), for instance, or to the need for codes of conduct that will guide scientists (Drenth 2006). In this discourse the basic perception of science is thus that it is, and should be, autonomous, but that scientists should both be reflexive about their practices and to some degree responsible for the outcomes of their scientific research upon society.

In contrast, the 'democracy discourse' argues that external actors need to be involved in the regulation and development of science and technology. In this discourse the basic problem is understood as being that new technologies and scientific developments are not sufficiently in line with public norms and values; such technologies and developments therefore need to come under a form of democratic control. This situation calls for the creation of structures that will enable external actors to be actively involved, for instance in constructive technology assessment (Schot and Rip 1997), ethical technology assessment (van der Burg 2009), or real time technology assessment (Guston and Sarewitz 2002). These processes are understood as functioning to align scientific development and research conduct with social and public values, although accounts differ in what they see as the central focus for change. Some are primarily concerned with the level of the individual scientist (van der Burg 2009; Scriebinger and Scraudner 2011), whereas others see it as crucial to change the scientific sector at the institutional or structural level (Schot and Rip 1997; but also Owen and Goldberg 2010; Owen et al. 2012; Swierstra and Jelsma 2006). Again, what unites these accounts as a single discourse is that external actors need to be explicitly involved in the regulation of science in order to secure its development according to shared social and political values. Scientists are not considered able to do this on their own: the entire process of science, then, from idea to outcome, is viewed as being integrated with a democratic structure that seeks to ensure socially accountable and desirable outcomes.

'Responsibility' and responsible development of science in the higher education, research ethics and sociology of science literature may, then, be articulated along a number of faultlines. Putting responsibility into practice can be construed as a process of separating science and society as far as possible (in the demarcation discourse), or of ensuring science's usefulness (in the usability discourse), or scientific reflection upon implications (the reflexivity discourse), or of enabling democratic deliberation and control of technoscience (the democracy discourse). Each of these will, of course, look somewhat different: the discourses take different sites and actors as their primary focus, and imagine responsibility as being centred around different kinds of actions (rigorous scientific method in the demarcation discourse, for instance, versus open public deliberation and debate in the democracy discourse). The key point we want to take up here is the multiplicity of 'responsibility'. Even the academic literatures that treat the notion do not present it consistently; we might anticipate, then, that there is similar diversity in 'real world' contexts in which the term is used. It is this question that we take up in the next section, visiting two empirical cases to outline the versions of 'responsibility' mobilised within them and their connection to the discourses described above.

9.3 Articulating Responsibility in Two US Sites

In what follows we discuss two examples of responsibility in practice, using these as instances of the way in which 'responsible development' can be very differently understood – and practiced – in different contexts. As we do this, we want to firmly locate ourselves within the discussion, acknowledging that we do not write as impartial observers – if such a thing were ever possible (Hammersley and Atkinson 1995) – but as participants in the data we will draw upon. Our analysis draws upon the story of a project, in which one of us – SRD – has been closely involved. This project was, for the period of 2010–2011, a central part of the activities of the National Science Foundation-funded Center for Nanotechnology in Society at Arizona State University (CNS-ASU), at which SRD was then based; her role included developing a programme of private sector engagement which would build the Center's private sector contacts and coordinate its outreach to and engagement with them. For CNS-ASU (and SRD), this private sector 'outreach' involved seeking to understand where CNS-ASU's points of contact and synergies with the private sector (interpreted broadly as including nano industry, business, not-for-profit research, and NGOs) lay, and developing these through informal conversations, community-building activities, and more sustained partnerships (see Davies 2011). Here, then, we are drawing upon SRD's situated and reflexive assessment of the progress of this outreach programme. Specifically, we will make use of a series of interviews she carried out with private sector actors around their understanding of responsibility; these more focused explorations are combined with a wider programme of participant observation (of, for instance, workshops on soft law, activities oriented around the status of the governance of nanotechnology, and local nano industry events) to build a picture of the way in which responsible development is being imagined within US-based private sector communities.

The two sites we will describe, then, are CNS-ASU and the private sector nanotechnology industrial actors with which it sought to interact. By using CNS-ASU's private sector engagement activities as the thread which ties these case studies of responsibility together, we are able to outline not just some of the multiplicities of the term but also the tensions which emerge when diverse articulations and practices around it come into contact. Our analysis is thus based not only on the ethnographic and interview data that SRD collected as she engaged in 'private sector outreach', but also on the ambiguities, frustrations, and hesitations – even, at times, discomfort – which were, for her, part of the lived experience of this project.

9.3.1 Responsibility in CNS-ASU: 'Real Time Technology Assessment' and 'Anticipatory Governance'

CNS-ASU does not, at first glance, frame itself in terms of enabling or ensuring the responsible development of nanotechnology. While the language it uses is somewhat different, its work can, however, be squarely placed within the broad moves around the governance of nanotechnology described in the introduction (Kearnes and Rip 2009), and its emergence as a key site of US thinking on societal and 'ELSI' (ethical, legal and social) implications has been concomitant with the development of research funding for nanotechnology more generally. Frameworks of 'anticipatory governance', 'real time technology assessment', and 'social and technical integration' developed within the Center have been influential within the wider governance landscape of both US and international nanotechnological development (Fisher 2007; Guston 2010, 2014; Guston and Sarewitz 2002; Kearnes and Rip 2009).

As David Guston, the Center Director, describes in a recent article (2010), CNS-ASU is funded by NSF from a wave of research monies made available by the National Nanotechnology Initiative (signed in 2000 and authorised in the 2003 twenty-first century Nanotechnology R&D Act). The NNI is of a piece with other research programmes for nanoscience and technology internationally in specifically legislating for the integration of social science, humanistic reflection, and public views with natural science research within the development of nanotechnology (Kearnes and Rip 2009): as Guston writes, the proposal for what became CNS-ASU was developed so as to 'address [the] congressional interest in public engagement with nano-scale science and engineering (NSE) research and the integration of societal concerns with NSE research' (Guston 2010, 432–433). The activities of CNS-ASU (and its sister center, at the University of California Santa Barbara) therefore play an important role in meeting the NNI's demands for a concern with 'societal dimensions' in the development of nanotechnology.

In responding to this broad remit CNS-ASU has worked with a toolkit of key concepts, including that of 'real time technology assessment' (Guston and Sarewitz 2002) – which provides a methodological focus for the Center's activities

(Guston 2010) – and, most significantly, anticipatory governance, a notion which, though it is increasingly being taken up within STS more widely (see Harvey and Salter 2012), remains a distinctive CNS-ASU 'brand' (Guston 2014). In the increasingly sophisticated accounts which are being given of anticipatory governance, its synergies with contemporary thinking on governance theory are emphasised (cf. Hajer and Wagenaar 2003), such that:

> anticipatory governance is meant to partake in a discussion in which governance is a wider set of activities than mere 'government' – that is, actions by public sector authorities. This wider set of activities is also meant to create between [sic] the two simple but extreme options that often dominate discussions of the governance of emerging technologies: 'doing' and 'banning.' […] Options between these two extremes include the implementation of licenses and other kinds of restrictions, the use of liability and indemnification, the application of intellectual property rights, the execution of treaties, the development of standards, testing regimes, and codes of conduct, and public action in various forms ranging from education to protest. (Guston 2010, 434)

The space in which anticipatory governance is operating is thus much the same as those in which discourses of responsible development are circulating. The 'options' Guston mentions – IPR, treaties and standards, codes of conduct, public action – largely overlap with those which are rolled in to discussions of responsibility as the key mode of governance of nanotechnology (Kearnes and Rip 2009; Kurath 2010). Anticipatory governance, however, is unusual in being both highly developed – in terms of the characterisation of what 'responsible' actions and activities should look like – and well articulated, with a small literature outlining its nature and practice (Barben et al. 2008; Davies and Selin 2012; Guston 2010, 2014). It is worth outlining its key dimensions as expressed in this literature.

Anticipatory governance, Guston writes, comprises 'four separate capacities that involve both research and practice: foresight, engagement, integration, and 'ensemble-ization'' (Guston 2010, 434). These capacities are *societal* (and thereby fit well with the NNI's emphasis on 'societal dimensions' of nanotechnology). They are envisaged as being developed within social science research institutions such as (and in particular) CNS-ASU, but it is fundamental to the way in which anticipatory governance is imagined that they extend beyond such research sites to incorporate, and affect, science (with a particular emphasis on the laboratory), policy, and society at large. Anticipatory governance therefore has a strong normative dimension in seeking to develop 'capacity for social learning' (Guston 2010, 433). 'Foresight' picks up on the notion of anticipation which, in this articulation, is explicitly framed not as prediction but as a means of imagining and deliberating multiple possible futures: techniques used include scenario workshops and the creation of physical prototypes. 'Engagement' has strong parallels with the now extensive literature on public engagement and deliberation on science (see, for instance, Hagendijk and Irwin 2006): it 'involves a connection between nanoscale science and engineering researchers with general publics that, at its best, provides for a two-way exchange of information and that tends to create a mutual understanding of values and goals' (Guston 2010, 435). 'Integration' focuses on engagements between natural scientists and social scientists and humanists, of which a paradigmatic example is 'social-technical integration research', in which graduate students from social science and humanities backgrounds spend time in natural science research labs in an effort to merge expertise from the different domains (Fisher 2007; Schuurbiers 2011). Finally, 'ensemble-ization' refers to the importance, within anticipatory governance, of none of these modes being understood as self-standing. In practice, Guston (2010) notes, successful CNS-ASU activities have tended to incorporate aspects of foresight, engagement, and integration.

CNS-ASU therefore has, in anticipatory governance, a highly developed model of what responsibility in nanotechnology could look like. With a budget of $6.5 million over a 5 year period, it also has scope – to an extent which is still unusual in social science research – to work on rolling out, at a relatively large scale, activities which develop these anticipatory capacities. Its programmes include lab-based 'integration' research, education of natural and social science university students, policy outreach, and a wide array of public engagement activities (see Davies et al. 2013; Davies and Selin 2012; Phelps and Fisher 2011; de Ridder-Vignone 2012; Selin and Boradkar 2010; Wender et al. 2012). While, as in any research centre, there is a degree of happenstance in what research is carried out, its highly focused structure (activities are organised according to what type of 'real time technology assessment', or RTTA, they fall under, and around a number of thematic research areas; Guston 2010) ensures that the different capacities of anticipatory governance are being developed in strategic and targeted ways. It is, within work at the Center, taken for granted that nanotechnology will best be developed through the heightening of these capacities. Notions such as societal engagement or the integration of social and technical expertise are therefore taken-for-granted principles within CNS-ASU contact with wider intellectual, political, and social worlds.

This, then, is the context from which CNS-ASU's private sector outreach was carried out – one which is saturated by a number of the discourses of responsibility described above. Anticipatory governance clearly takes as its starting point the assumption that the relationship between science and society should be strengthened and developed; nanotechnology, we are told, 'occasions new approaches to the conduct of research evaluation and assessment that require the engagement of a variety of potential users and stakeholders' (Barben et al. 2008, 979). Aspects of the reflexivity and democracy discourses are particularly strong. CNS-ASU is largely structured around enabling public action and debate around nanotechnology and ensuring that the reflexivity of science and scientists is enhanced through 'integration' activities.

How has anticipatory governance become what it is? While the genealogy claimed for the concept includes the environmental studies literature (Karinen and Guston 2010) and Alvin Toffler's 1971 book *Future Shock* (Guston 2010), it also seems clear that CNS-ASU's activities are – as with those of CBEN and ICON, in McCarthy and Kelty's study (2010) – of their time, and have been constituted through a variety of academic, political, and financial dynamics. CNS-ASU's work, as expressed in anticipatory governance, can be understood as a pragmatic response to the NNI's emphasis on the societal dimensions of nanotechnology, as well as to a wider intellectual climate in which public engagement and the 'democratisation of science' are important themes (see Hagendijk and Irwin 2006; Jasanoff 2003) and where there has been an international shift towards the governance – rather than government – of technology (Hajer and Wagenaar 2003). Just as the natural scientists of CBEN learned to deal with 'responsibility' in ways that protected both publics and their science (McCarthy and Kelty 2010), anticipatory governance can be seen as a strategic version of responsibility which has traction on the NNI – and thereby NSF funding – as well as maintaining rigour in the context of contemporary thought in STS and science policy.

9.3.2   *Responsibility in US Private Sector Nanotechnology: Safety and 'Not Blowing Things Up'*

For CNS-ASU, then, the responsible development of nanotechnology will inevitably involve attention to its wider societal dimensions, and will involve notions of democratic engagement and self-reflexive scientific practice. This is, of course, of a piece with many formal articulations of responsibility. As a number of scholars have charted (Kearnes and Rip 2009; Kurath 2010; Randles 2008), the 'governance landscape' of nanotechnology integrates calls for responsibility with activities such as soft law, codes of conduct, ethics research, and public engagement. Kearnes and Rip write that this:

> emerging landscape incorporates a number of recent regulatory initiatives, including: regulatory reviews concerning the sufficiency of existing regulatory frameworks for novel nanomaterials; the incorporation of ELSA research and public deliberation at upstream stages in the development of nanotechnology; the recent proliferation of principle-based voluntary codes concerning the development of nanotechnology and the broader emergence of a discourse of the 'responsible development of nanotechnology'. (Kearnes and Rip 2009)

There is thus significant overlap between 'responsibility' and – in particular – societal activities such as 'public deliberation'. Both the EC-CoC (see European Commission 2008; Kjølberg and Strand 2011) and the Responsible Nano Code (2008), for example, incorporate principles such as accountability, inclusiveness (that 'governance of N&N research activities should be guided by the principles of openness to all stakeholders'; European Commission 2008, 6), stakeholder involvement, and transparency.

CNS-ASU is not so unusual, then, in mobilising this rather broad model of responsibility. The story we want to tell in this section, however, is that this stands in stark contrast to the version of responsibility assumed by the US private sector organisations and individuals with whom CNS-ASU has had contact in its outreach programme. For US nano industry – as encapsulated by the community with which SRD engaged through attendance at nano-oriented conferences and meetings, and through a set of interviews and informal conversations with key informants – responsibility is almost entirely constituted around being safe. Responsible development is thus seen as being about 'not blowing things up', worker safety, risk management, environmental health and safety (EHS), and the 'protection of public health and the environment': there is little, if any, discussion of broader questions such as the direction that technology should be taking.

This can perhaps best be illustrated by giving some instances of responses, within interviews, to the question: what does responsible development of nanotechnology mean to you? The answers given almost exclusively emphasised notions such as the management of toxicity, risks and benefits (and calculative approaches to dealing with these), and the development of new testing procedures. The extracts below are indicative.

> Yeah, I think um nanotechnologies, and really new chemical developments in general, represents a real opportunity for green chemistry, and what's called benign by design. So the idea that you could do your toxicity testing early in the chemical development and in the application of that chemical um and understanding the toxicology or the potential hazards of that material you might be able to design in the R&D phase to make it safer or less risky. (Interview 1, NGO representative)

> So it's a weighing of risks and benefits, and I guess ultimately that to me is what responsible development means, is don't look just at the risks, don't look just at the benefits, look at both and try to assess them and try to um you know maximize the benefits and minimize the risks. But people will have different calculuses for exactly what decisions should be made in that type of framework. (Interview 2, Industry representative)

In the first quote the speaker – who is from an environmental NGO – emphasises nanotechnology's potential for enabling so-called 'benign by design' chemicals. In so doing she picks up on discourses in which control of matter, at the nanoscale, will enable control of material effects: safety, in the form of an awareness of 'toxicology or the potential hazards' can be engineered into new materials such that they can be made 'less risky' (cf. Schwarz 2009). The notion of risk is similarly key in the second extract. The immediate response to the nature of responsible development is that it is fundamentally a 'weighing of risks and benefits', and thereby a 'calculus' by which nanotechnology's benefits and hazards can be assessed. While the speaker acknowledges that such calculations may vary from person to person (that 'people will have different calculuses'), the assumption is that risk and benefit are the only categories at stake (cf. Kearnes et al. 2006). In both instances, then, the interviewees turn directly to questions of risk as the central meaning of responsibility.

This reading of responsible development as primarily about ensuring (relative) safety was present throughout the interviews and the wider engagement activities of CNS-ASU's private sector outreach. A meeting on soft law with a large industry audience, for instance, which SRD attended with some expectation of hearing about the wider societal engagements outlined in various Codes of Conduct and proposed within anticipatory governance, was dominated by discussion of risk analysis and management. The quote that heads this section – taken from a speaker's comment that their primary concern was simply 'not blowing things up' – similarly comes from a local nano industry meeting, at which the implications of a particular technological application were discussed. And within the interviews and other informal conversations as a whole there was little sense that responsible development might, or could, ever accommodate the wider notions of societal engagement, public debate, scientific and social integration, and anticipation and foresight which are rolled into other versions of responsibility (cf. Kjølberg and Strand 2011).

In this context, then, a much narrower version of responsibility is being mobilised – one which construes it as essentially about the management of (physical) risk and which depicts two key dichotomies: safety versus risk; and risk versus bene-fit. The 'responsible' response to these dynamics is to seek to control and manage them; hence, in interviews and discussion, the emphasis on technical methods and procedures to limit or understand risk and uncertainty (see Dupont and EDF 2007; cf. Kearnes et al. 2006). As such, there are clear parallels with the *reflexivity* discourse of responsibility outlined above – though here reflexivity is limited to physical harm rather than any wider outcomes or effects on society, and is therefore a rather different kind of reflection than that promoted by CNS-ASU (Fisher 2007). While we can also observe traces of the usability discourse – in the assumption that responsibility is enacted within marketplace dynamics – the discourse which CNS-ASU makes primary use of, that of democratisation, is strikingly absent from private sector discussion.

The purpose of this assessment is not to criticise the version of responsibility mobilised by private sector actors, or to hold it up to some standard of which antici-patory governance is the benchmark. Just as we earlier argued that anticipatory governance – as expressed within the activities and foci of CNS-ASU – is a product of particular dynamics operating at a particular time, the same is equally true of US private sector nanotechnology. If CNS-ASU and its development of frameworks such as sociotechnical integration, real time technology assessment, and anticipa-tory governance can be understood as in part shaped by political moves which both promote nanotechnology and demand that it is socially aware, then private sector discourses of responsibility which solely emphasise risk-based approaches are simi-larly responses to a particular kind of context. At one level, of course, this context is the same: the same macro level demands for 'responsible development' are operating (as expressed, for instance, in increased discussion of corporate social responsibility and the ways in which this applies to nanotechnology; see Groves et al. 2011; Krupp and Holliday 2005). But our engagement with private sector actors indicates that there at least two other factors which shape industry articula-tions of responsibility: the logic of the bottom line, and the practical limitations of actors' agency and social milieu.

Neither of these are, of course, surprising. The private sector is, largely, 'about making money', as one informant noted, and the maximisation of profits will inevi-tably be a key filter through which any broader demands will be viewed. Thus, for this community, responsibility tends to be articulated in ways which fit with this goal. While it is easy to trace a connection between risk management – 'doing no harm' – and financial stability, it is less so for wider imaginations of responsibility such as societal debate, transparency, and sociotechnical integration (cf. Groves et al. 2011). Beyond this, however, it was also clear that many of our informants had a strong awareness of the opportunities and constraints of the situations they – as individuals, businesses, and whole industries – were in. They were experts in a specific domain – whether nano-related law, environmental policy, or nanotech start-ups – and their perspectives were shaped by these domains. Responsibility was thus almost always seen in the context of a relatively narrow field of action and opportunity – for the speaker below, for instance, it came down to a particular way of doing chemistry.

> I think when it comes to the responsible area- one of the issues is that each of the sciences
> has its weak spot, we in the colloid chemistry area do not really define our surfaces very
> well. […] So going back to basics, re-examining the purpose of the test, making certain that
> you're using the best evaluation technique, and each of the disciplines shall we say stepping
> up… (Interview 5, Industry representative).

Colloid chemistry, tests, and evaluation techniques are what this speaker knows, in astonishing detail and from some decades of working in this field; for them, then, responsible development has to be something which can be articulated in these terms. Indeed, for responsibility to be made meaningful to any of our informants, it had to be something that they were able to act upon (however weakly) within the confines and constraints of their immediate context. Broader visions of directing 'nanotechnology' as a whole, enabling societal debate, or anticipating disruptive technologies were, largely, simply not possible. For responsible development to be 'made do-able' – to use McCarthy and Kelty's term (2010) – it was therefore necessary to strip it to the bare bones of risk assessment and cost-benefit analysis – to make it something that our informants, and private sector nano more generally, could mobilise their expertise to act upon.

9.4   Conclusion

In this essay we have been concerned with the multiplicity of responsibility. Our argument has been that, though it is easy to take calls for responsible development of technology for granted, there is much more going on underneath the surface of such calls than we might anticipate: though there is certainly homogeneity around the language of responsibility – with it having become, particularly in the context of nanotechnology, a standard framework for technological development – its meaning is contingent and uncertain. In academic literature that treats responsibility in science, for instance, we can distinguish a number of discourses which outline what such responsibility should look like. While all of these focus on problematic features of the science-society relationship, they suggest different diagnoses and remedies, from complete demarcation to enhancing usability, reflexivity or democracy. As such, these discourses might be understood as 'ideal types' (cf. Weber 1949) of responsibility – purified forms of the different ways in which the term can be understood and performed.

In contrast, we have found that the ways in which responsibility is articulated in 'real world' sites is messy and contingent, shaped by dynamics and pressures which may be connected to the politics of responsibility (the widely shared imagination of nanotechnology as an opportunity to 'get things right') but which also may not (the need for companies to be mindful of the bottom line). The versions of responsibility which are mobilised in CNS-ASU and in the US nanotechnology private sector do not fall neatly along the lines outlined by the different discourses – or types – the literature discusses; rather, what we see is a combination of the themes around the sites and purposes of responsibility which emerge in more formalised articulations. Thus CNS-ASU, through its model of anticipatory governance, brings together an inclusive picture of responsibility which incorporates an emphasis on democracy – in the shape of societal capacity for anticipation – and on reflexivity (through real time technology assessment and lab-based reflection). In contrast, the nanotechnology industry appears to rely on narrower versions of reflexivity which focus on safety and which give little attention to wider public debate.

In all of this the key point is that responsibility is various. It is various in the literature, and it is certainly various in the ways it is performed within different sites. Perhaps most critically, it is various in ways that may at first glance be hidden or disguised: the very uniformity of the language of responsible development, and the solidity of assumptions of what it is, can mean that there is little explicit articulation of what it means in practice. The project that we have described – of CNS-ASU private sector outreach – is a prime example of this. CNS-ASU, as we have noted, takes for granted an expansive version of responsibility in line with policy such as the EC-CoC; for this to come into contact with the much more tightly understood version of responsibility – in which it entails safety, and not much more – espoused by the private sector was at times disruptive for both CNS-ASU and industry actors. The point with which we want to close, then, echoes that made by others who have started to unpick the multivalencies of 'responsibility' (Kjølberg and Strand 2011), and is simply that to talk only of responsibility, or responsible development, may be worse than useless. The key question is of what *kind* – what version – of responsibility we are calling for, enacting, and transporting as we study, and write about, technological development.

References

Barben, D., E. Fisher, C. Selin, and D.H. Guston. 2008. Anticipatory governance of nanotechnology: Foresight, engagement, and integration. In *The handbook of science and technology studies*, 3rd ed, ed. E.J. Hackett, O. Amsterdamska, M. Lynch, and J. Wajcman, 979–1000. Cambridge, MA: MIT Press.

Barbosa, N., and A.P. Faria. 2011. Innovation across Europe. How important are institutional differences? *Research Policy* 40(9): 1157–1169.

Corolleur, C.D.F., M. Carrere, and V. Mangematin. 2004. Turning scientific knowledge into capital. The experience of biotech startups in France. *Research Policy* 33(4): 631–642.

Cross, R.T., and R.F. Price. 1999. The social responsibility of science and the public understanding of science. *International Journal of Science Education* 21(7): 775–785.

Davies, S.R. 2011. *Nanotechnology, business, and anticipatory governance*. CNS-ASU Report #R11-0004, Center for Nanotechnology in Society. Arizona State University. Tempe, AZ.

Davies, S.R., and C. Selin. 2012. Energy futures: Five dilemmas of the practice of anticipatory governance. *Environmental Communication: A Journal of Nature and Culture* 6(1): 119–136.

Davies, S.R., C. Selin, G. Gano, and A. Guimarães Pereira. 2013. Finding futures: A spatio-visual experiment in participatory engagement. *Leonardo* 46(1): 76–77.

Deichmann, U. 2005. Unholy alliances. *Nature* 405(6788): 739.

de Ridder-Vignone, K. 2012. Public engagement and the art of nanotechnology. *Leonardo* 45(5): 433–438.

Drenth, P.J.D. 2006. Responsible conduct in research. *Science and Engineering Ethics* 12(1): 13–21.

Dupont and EDF. 2007. *Nano risk framework*. Washington, DC: Environmental Defense – Dupont Nano Partnership.

Durant, J., G. Evans, and G.P. Thomas. 1989. The public understanding of science. *Nature* 340: 11–14.

European Commission. 2004. *Towards a European strategy for nanotechnology*. Luxembourg: Commission of the European Communities.

European Commission. 2008. *Commission recommendation of 07/02/2008 on a code of conduct for responsible nanosciences and nanotechnologies research*. Brussels: European Commission.

Fisher, E. 2007. Ethnographic invention: Probing the capacity of laboratory decisions. *NanoEthics* 1(2): 155–165.

Forge, J. 2000. Moral responsibility and the 'ignorant scientist'. *Science and Engineering Ethics* 6(3): 341–349.

Gilmer, P.J., and M. DuBois. 2002. Teaching social responsibility: The Manhattan Project: Commentary on "the six domains of research". *Science and Engineering Ethics* 8(2): 206–210.

Glerup, Cecilie, and Maja Horst. 2014. Mapping 'social responsibility' in science. *Journal of Responsible Innovation*: 1–20. doi:10.1080/23299460.2014.882077

Groves, C., F. Lori, R. Lee, and E. Stokes. 2011. Is there room at the bottom for CSR? Corporate social responsibility and nanotechnology in the UK. *Journal of Business Ethics* 101(4): 525–552.

Guston, D.H. 2010. The anticipatory governance of emerging technologies. *Journal of the Korean Vacuum Society* 19(6): 432–441.

Guston, D.H. 2014. Understanding 'anticipatory Governance.' *Social Studies of Science* 44(2): 218–242. doi:10.1177/0306312713508669.

Guston, D.H., and D. Sarewitz. 2002. Real-time technology assessment. *Technology in Society* 24(1–2): 93–109.

Hagendijk, R., and A. Irwin. 2006. Public deliberation and governance: Engaging with science and technology in contemporary Europe. *Minerva* 44(2): 167–184.

Hajer, M.A., and H. Wagenaar. 2003. *Deliberative policy analysis: Understanding governance in the network society*. Cambridge: Cambridge University Press.

Hammersley, M., and P. Atkinson. 1995. *Ethnography: Principles in practice*. London/New York: Routledge.

Harvey, A., and B. Salter. 2012. Anticipatory governance: Bioethical expertise for human/animal chimeras. *Science as Culture* 21(3): 1–23.

Jasanoff, S. 2003. Technologies of humility: Citizen participation in governing science. *Minerva* 41: 223–244.

Jones, R. 2008. When it pays to ask the public. *Nature Nanotechnology* 3(10): 578–579.

Karinen, R., and D.H. Guston. 2010. Toward anticipatory governance: The experience with nanotechnology. In *Governing future technologies: Nanotechnology and the rise of an assessment regime*, ed. M. Kaiser, M. Kurath, S. Maasen, and C. Rehmann-Sutter, 217–232. Dordrecht: Springer.

Kearnes, M.B., and A. Rip. 2009. The emerging governance landscape of nanotechnology. In *Jenseits Von Regulierung: Zum Politischen Umgang Mit Der Nanotechnologie*, ed. S. Gammel, A. Losch, and A. Nordmann, 97–121. Berlin: Akademische Verlagsgesellschaft.

Kearnes, M.B., P. Macnaghten, and J. Wilsdon. 2006. *Governing at the nanoscale: People, policies and emerging technologies*. London: Demos.

Kjølberg, K.L., and R. Strand. 2011. Conversations about responsible nanoresearch. *NanoEthics* 5(1): 99–113.

Kowal, J.P. 1980. Responsible science reporting in a technological age. *Journal of Technical Writing and Communication* 10(4): 307–314.

Krupp, F., and C. Holliday. 2005. Let's get nanotech right. *The Wall Street Journal*, June 14, p. B2.

Kurath, M. 2010. Nanotechnology governance. *Science, Technology, and Innovation Studies* 5(2): 87–110.

Leydesdorff, L., and H. Etzkowitz. 1998. The triple helix as a model for innovation studies. *Science and Public Policy* 25(3): 195–203.

Macnaghten, P., M.B. Kearnes, and B. Wynne. 2005. Nanotechnology, governance, and public deliberation: What role for the social sciences? *Science Communication* 27(2): 268–291.

McCarthy, E., and C. Kelty. 2010. Responsibility and nanotechnology. *Social Studies of Science* 40(3): 405–432.

Merton, R.K. 1973. The normative structure of science. In *The sociology of science: Theoretical and empirical investigations*, ed. R.K. Merton, 223–266. Chicago: University of Chicago Press.

Michael, M. 2002. Comprehension, apprehension, prehension: Heterogeneity and the public understanding of science. *Science, Technology, and Human Values* 27(3): 357–378.

National Nanotechnology Initiative. n.d. *National nanotechnology initiative.* http://www.nano.gov/nanotech-101/what/definition. Accessed 20 Oct 2012.

Owen, R., and N. Goldberg. 2010. Responsible innovation: A pilot study with the UK Engineering and Physical Sciences Research Council. *Risk Analysis* 30(11): 1699–1707.

Owen, R., P. Macnaghten, and J. Stilgoe. 2012. Responsible research and innovation: From science in society to science for society, with society. *Science and Public Policy* 39(6): 751–760.

Phelps, R., and E. Fisher. 2011. Legislating the laboratory? Promotion and precaution in a nano-materials company. *Biomedical Nanotechnology* 726: 339–358.

Randles, S. 2008. From nano-ethicswash to real-time regulation. *Journal of Industrial Ecology* 12(3): 270–274.

Responsible Nano Code. 2008. *Information on the responsible nano code initiative.* London: Responsible Futures.

Roco, M.C., B. Harthorn, D. Guston, and P. Shapira. 2011. Innovative and responsible governance of nanotechnology for societal development. *Journal of Nanoparticle Research* 13(9): 3557–3590.

Schuurbiers, D. 2011. What happens in the lab: Applying midstream modulation to enhance critical reflection in the laboratory. *Science and Engineering Ethics* 17(4): 769–788.

Schwarz, A. 2009. Green dreams of reason. Green nanotechnology between visions of excess and control. *NanoEthics* 3(2): 109–118.

Schot, J., and A. Rip. 1997. The past and future of constructive technology assessment. *Technological Forecasting and Social Change* 54: 251–268.

Scriebinger, L., and M. Scraudner. 2011. Interdisciplinary approaches to achieving gendered innovations in science, medicine, and engineering. *Interdisciplinary Science Reviews* 36(2): 154–167.

Selin, C., and P. Boradkar. 2010. Prototyping nanotechnology: A transdisciplinary approach to responsible innovation. *Journal of Nano Education* 1(1–2): 1–12.

Shelley-Egan, C. 2010. The ambivalence of promising technology. *NanoEthics* 4(2): 183–189.

Shelley-Egan, C., and S.R. Davies. 2013. Nano-industry operationalizations of 'responsibility': Charting diversity in the enactment of responsibility. *Review of Policy Research* 30(5): 588–604.

Stilgoe, J., R. Owen, and P. Macnaghten. 2013. Developing a framework for responsible innovation. *Research Policy* 42(9): 1568–1580.

Swierstra, T., and J. Jelsma. 2006. Responsibility without moralism in technoscientific design practice. *Science, Technology & Human Values* 31(3): 309–332.

van der Burg, S. 2009. Taking the "soft impacts" of technology into account: Broadening the discourse in research practice. *Social Epistemology: A Journal of Knowledge, Culture and Policy* 23(3–4).

Weber, Max. 1949. *On the methodology of the social sciences.* Glencoe: The Free Press.

Wender, B.A., R.W. Foley, D.H. Guston, T.P. Seager, and A. Wiek. 2012. Anticipatory governance and anticipatory life cycle assessment of single wall carbon nanotube anode lithium ion batteries. *Nanotechnology, Law and Business* 9(3): 101–118.

Wolpert, L. 2005. The Medawar Lecture 1998 is science dangerous? Philosophical transactions. *Royal Society. Biological Sciences* 360(1458): 1253–1258.

Chapter 10
Nanotechnology and Configurations of Responsibilities in Boundary Organizations

Paolo Magaudda

10.1   Introduction: Responsibility and New Organizational Configurations in Nanotechnology

In the last few years, the debate about advances in nanotechnology and the social and political responsibility implied in these advances has grown among scholars and politicians (Robinson 2009; McCarthy and Kelty 2010). The discussion about responsibility connected with nanotechnological innovations has developed in different ways. Firstly, in the last decade, nanotechnologies have played the role of 'frontier' research field, attracting huge research funds and raising public interests in the media and popular culture (Neresini 2006; Nerlich 2008). Moreover, debates about nanotechnology have been characterised by a recurring perceived distance between, on the one hand, the great theoretical promises and, on the other hand, the mundane practical applications actually developed, leaving substantial space to reflect on the possible forthcoming consequences of nanotechnologies, maybe even unexpected and potentially dangerous ones (Selin 2007; Bijker et al. 2010). Moreover, we should not forget that these concerns around nanotechnology touch very sensitive areas for human society, such as in the case of the effect on the environment and on human health and bodies (Roco 2005; Faunce 2007). Clearly, the importance of nanotechnology's possible consequences has spurred the growth of reflections and analyses about the taking of responsibility for positive and negative changes produced or inspired, directly or indirectly, by nanotechnological research (Roco and Baingridge 2001).

Going beyond the reasons why the debate on the link between responsibility and nanotechnology is going to be increasingly relevant, there is another important perspective to add to this discussion, which until now has been scarcely considered.

P. Magaudda (✉)
Department of Philosophy, Sociology, Education and Applied Psychology,
University of Padova, Padova, Italy
e-mail: paolo.magaudda@unipd.it

S. Arnaldi et al. (eds.), *Responsibility in Nanotechnology Development*, The International Library of Ethics, Law and Technology 13, DOI 10.1007/978-94-017-9103-8_10,
© Springer Science+Business Media Dordrecht 2014

The debate on the responsibility for nanotechnology innovation has taken into account only in a very superficial way how nanotechnologies have required and also supported changes in the organizational configurations of actors and institutions involved in the development of this scientific and economic field (Robinson 2009; Rampersad et al. 2010). The pressure for new organizational arrangements in scientific practices has laid the foundation for the development of renewed models of innovation, different from more traditional ways to manage innovation's processes and actors. When looking closer at this issue, we can see that the request for more flexible relationships between market, state and scientific research have impacted not only the institutional relationships between actors, but also the ways they frame their reciprocal responsibilities (see Etzkowitz and Leydesdorff 2000). It is not hard to speculate about how the development of new relationships between actors in nanotechnology has also produced effects on the framing of responsibility in the innovation processes and practices.

In fact, the change toward new organizational configurations in innovation work has acquired significance especially in the nanotechnological field. Organizational changes have gone in the direction of a more complex relationship between market actors and scientific institutions, leading to the increasing importance of the creation and the maintenance of stable relationships between different spheres, interests and actors. Among the several ways to describe these new configurations in the field of innovation, and their resulting tensions, a useful perspective is that developed around the concept of "boundary organizations" (Guston 1999, 2001). By using this concept, it has been possible to pay attention to the relevance of more flexible and smooth divisions of tasks, activities and responsibilities in innovation paths. Moreover, the focus on the boundary dimension of cooperation has been exemplified by the increasing growth of forms of 'hybrid management' in research and development (Miller 2001). In the context of the contemporary changes of scientific work, 'boundary organizations' and 'hybrid managements' represent useful insights to be added to the evolving debate concerning the interaction between responsibility and nanotechnology.

In this chapter I aim to widen the understanding of the multiple analytical levels of the relationship between responsibility and nanotechnology, paying specific attention to the changing organizational configurations in the cooperation between different actors involved in nanotechnology innovation. More specifically, I start by asking: how the emergence of new and more flexible organizational configurations within the nanotechnology field is entangled with evolving forms of responsibility; how responsibility is actually addressed by different actors involved in the processes, and; whether these more flexible relationships between state, market and research actors are promoting or rather hindering new configurations of responsibilities.

This chapter develops some of these issues by considering the history and activities of a Italian 'boundary organization' in the field of nanotechnology, paying particular attention to the ways new organizational configurations have implications on the emerging configuration of responsibility. More specifically, this chapter focuses on some of the aspects involved in the development of a small nanotechnology facility characterized by the direct interaction between firms, managers, entrepreneurs and scientific researchers. By reconstructing the pattern of creation and the work developed within this nanotechnological facility, the chapter aims at highlighting how its organizational shape is connected with the emergence of new practical and discursive configurations of responsibility. The chapter develops some considerations starting from a case-study based on data generated with qualitative interviews and document analysis and, points out the 'heterogeneousness' which today characterizes the work of innovation (Law 1987; Latour and Woolgar 1979; Knorr-Cetina 1981). In addition, the chapter also looks at how innovation involves a multiple set of actors, contexts and players of broader socio-cultural backgrounds, that are locally embedded and, thus also opened to the influences coming from different interests and perspectives (Pinch and Bijker 1984; Latour 1987; Law and Hassard 1999). This approach allows an understanding the several levels of responsibility and how these levels are articulated in the actual contexts of innovation. It also helps us to figure out how innovation paths in nanotechnology emerge from a work of 'co-production' played by different actors rooted in specific local contexts (Doubleday 2007). More generally, the analysis developed in this chapter contributes to further the debate on the relationship between nanotechnology and society, by generating a 'greater reflexive awareness among scientists in their specialist work worlds, with the expected result that innovation processes indirectly gain added sensitivity to human needs and aspirations, and thus greater resilience and sustainability' (Macnaghten et al. 2005, 271).

Having these analytical questions in mind, I intend to explore empirically the work of Nanofab, an Italian nanotechnology laboratory located outside Venice. The research on which the chapter is based has been carried out for about 1 year between January 2007 and March 2008 and relies on an empirical point of view on ethnographic observations, document analysis and in-depth interviews with the staff of Nanofab, including both those in purely scientific roles (project manager and researchers) and those in managerial, administrative and advisory positions.

In the first part, the chapter describes the start of Nanofab and the beginning of the construction of its laboratories located in Marghera, an area on the outskirts of Venice. Nanofab started around 2000 and its story is told in the chapter paying particular attention to how the design of the project required a set of negotiations between scientific requirements, institutional activities and constraints posed by institutional public actors. Afterwards, the chapter shows how the cooperation to transfer nanotechnologies to the business environment posed a number of problems to those actors and institutions involved. These problems affected, at the same time, the scientific and organizational structure, the interactions between actors and, consequently, the framework within which forms of responsibility have been attributed and agreed upon. The chapter explores these issues by focussing on the actual ways in which managers, scientists and businessmen interacted to establish forms of cooperation and to share reciprocal responsibility and trust. Finally, in the conclusion the chapter discusses how the case of Nanofab represents a useful and fruitful point of departure to widen our understanding of the relationship between responsibility and nanotechnology and to feed the on-going debate regarding different forms of responsibility in real terms in the work of nanotechnology innovation.

10.2   New Configurations in Nanotechnology Innovation: The Nanofab Case

Nanofab is a not-for-profit organization set up to develop and help the transferring of nanotechnology-based processes and knowledge from the academic sector to local industries, largely characterised by being small and middle-sized firms. Nanofab's activities could thus be rightly considered as belonging to the realm of technology transfer from the scientific sector to firms. The institutional actor who promoted Nanofab is the Veneto region, and among the main institutional supporters there are at least two important players. The first one is Veneto Nanotech, a regional based public agency which coordinates regional activities about nanotechnology; the other is CIVEN, an academic network including the four main universities in the region (the Universities of Padova, of Verona and of Venice-Ca'Foscari and – at a later stage – Venice IUAV) offering the scientific and technical know how to Nanofab. The academic network CIVEN plays the role of a sort of tank from which Nanofab is able to draw innovative knowledge and highly specialized skills to be applied in the transfer of nanotechnology. Another relevant actor also involved in Nanofab is a science park named Vega, located on the Venice mainland, where Nanofab has now its own headquarter and laboratories. The main scientific fields covered by Nanofab include nanostructured materials and thin films, nanopowders, and in general all those applications aimed at improving the performances of material surfaces in products and tools.

Nanofab's mission concerns the transfer of technological knowledge to the market, including industrial testing, in collaboration with industries and specifically the small and medium-sized businesses located in the highly industrialized area of the Veneto Region. The local industrial sectors most interested in applying nanotechnology-based innovations are those of the clothing and footwear industries, jewellery, the fashion industry, machinery, and also rubber-, plastics- and leather-related producers of consumer goods. Moreover, we should note that Nanofab fits into a larger national strategy based on the establishment of a technological cluster for nanotechnology in the Veneto region, which started in 2004. As we will now show, the birth of Nanofab in 2005 represented the result of a synergistic interaction between scientific institutions, public and institutional actors with small and specialized firms in order to introduce nanotechnologies in the market.

Nanofab's activities started in October 2005, when the headquarters and laboratories were officially opened. However, the process of planning and networking necessary for the creation of Nanofab originated a few years earlier. In fact, as is always the case when we focus on the development of complex socio-technical processes, the trajectory that goes from the initial idea to the actual creation of a new research structure is intricate and highly differentiated. As one of the leading scientific promoters of the project told us, the idea of creating Nanofab took shape at least 5 years before its actual opening, around the year 2000. At this early stage, a group of Italian chemical scientists organized a trip to the U.S. in order to visit the evolving businesses related to the then new-born National Nanotechnology Initiative (NNI), the US research program aimed also at developing new forms of cooperation between the academia and private firms in the field of nanotechnology:

> One of these centres [we visited] was closer to the model we later used for Nanofab; it was the Pennsylvania State University. [...] And the Pennsylvania State University had established an entity called the Nanofabrication Facility, which was at the service of companies, but also [suitable] for academic activities [...]. And during the trip the idea that perhaps the Veneto region could finance an initiative in nanotechnology like that one finally arose (Interview no. 5).

After getting their initial idea from the organizational developments happening in the US, the original group of scientists started to 'enrol' the institutional actors within and outside the academy with the goal of developing a similar project tailored to the needs (and limits) of the Veneto region's small firms and concentrated industrial area. In 2002, the Veneto region finally financed the creation of new laboratories inside the Vega science park in Marghera, involving more directly scientific expertise located within the various universities in the regions with research activities the nanotechnology field, grouped together into the CIVEN consortium.

The construction of Nanofab's laboratories required the recruitment of both political actors and institutional resources necessary to ensure sound economic and structural foundations to the project. This process of enrolment was settled together with the configuration of responsibilities between different partners in the network. The institutional actors of the network assumed the role of sponsorship of the project with the regional industrial clusters, while scholars and academic institutions took the responsibility for the development of Nanofab in terms of technical equipment and scientific guidance.

The Nanofab project developed an emerging configuration of responsibility together, which has been characterized by the intersection of several logics, or perspectives. These logics have been in part related with the needs settled by techno-scientific problems and in part they have arisen as consequence of a series of institutional, materials and economic constraints. For example, while from a scientific point of view the natural location for Nanofab's laboratories would be the city of Padua, where the highest academic and scientific resources in the field of nanotechnology are to be found, the definitive choice for the placement of the laboratories was the Venice mainland area. This decision was taken on the basis of administrative and financial opportunities recognized at institutional level by the Veneto region. This was described by one of the Nanofab's founders thus:

> Why Venice? Why did this thing happened in Venice, when no-one would question that the technical and scientific expertise in this field is in Padua? Because the funds used were coming from the European regional development fund and had to be used in the Marghera area (Interview no. 5).

A further constraint in setting up Nanofab regarded the need to identify the needs of local small firms intended as the primary beneficiaries of the transfer of nanotechnology. Thus, the construction of the network of innovation required taking into account also the responsibility of public institutions in terms of local development and contextual requests negotiated among the different local stakeholders involved.

For example, one of the main aspects influencing the definition of scientific areas to be developed within Nanofab required a negotiation between scientific experts from the academy, the political interests coming from the Veneto region and the commercial interests in the regional entrepreneurial setting:

> This constraint concerning the fact that we were going to make a technological cluster in a very advanced area of scientific and technological developments, but able also produce results on the market, on processes and products, has been very important. If we had made this choice directly, […] we would have lost the industrial support. Because industrial firms could not wait five or ten years. With this decision, of course, we were conscious that we were creating a structure suitable more for businesses than to academic research (Interview no. 5).

While the original idea of Nanofab has born in a scientific context, inspired by similar international experiences occurring in the field of nanotechnology, the establishment of an actual network rooted in the local industrial environment introduced the need for a set of negotiations regarding both institutional and scientific choices. The same configuration of actors involved reflected and sustained the emergence of a specific framework of responsibilities, in which local institutions identified specific responsibilities toward the regional firm industrial cluster existing in the Veneto region.

During the process of Nanofab's design, a crucial passage was the creation at national level of the Italian technology clusters (as happened between 2002 and 2004) and the selection of the Veneto region as national reference cluster for nanotechnology. These Italian clusters are areas geographically defined usually at regional level, characterized by a clear expertise in specific scientific and industrial sectors such as biotechnology, ICT and nanotechnology (see Miceli 2010). The decision to establish the cluster for nanotechnology in the Veneto Region offered a more solid framework for the players involved in Nanofab. Another important passage in the creation of the Nanofab network was the establishment of CIVEN – the academic consortium holding scientific expertise about nanotechnology – that emerged as a common effort from the cooperation of the various universities located in the Veneto area. As told by a manager of the Veneto cluster, the collaboration between universities present in the project was a key impulse for the successful development of Nanofab and implied a distribution of commitments and responsibilities between universities and between regional institutions and academic structures:

> CIVEN was created precisely because the region urged the three universities in the Veneto to give an answer in terms of projects of this type, and actually – it's rather rare case – the universities responded in a unique way by creating a single entity that is a combination of the three universities and to which is partially delegated the management of the cluster (Interview no. 1).

All those involved in the CIVEN emphasize that the collaboration between universities has represented a valuable add-on, because it made it possible, as noted by another leading scientists of the project, to overcome the particular interests of the academic centres involved. Especially in the initial period of Nanofab's existence, the inter-universities consortium constituted an essential resource for the scientific activities, particularly because CIVEN's researchers had direct responsibility in the creation of Nanofab's laboratories.

On the one hand, at the beginning we started working together especially with Padua and Venice, while Verona came later. It worked well basically, there were no tensions or, you know, special interests. And then, for better or worse, from the Veneto region there was the idea: 'Why do not we form a single entity?'. And it was, I guess you could say, a winning move (Interview no. 2).

These short insights into the setting up of this project allow us to focus more carefully on the idea that the process of establishing Nanofab required the mobilization of several different interests, the configuration of responsibilities and the subsequent large-scale cooperation at both scientific and political level. In this sense, Nanofab has emerged as the dynamic output of a complex of set of relationships, interactions and mutual responsibilities between the actors and institutions enrolled into the project.

10.3 Innovation and Responsibility in Nanotechnology Boundary Organizations

One of the main metaphors used in the last decade to account for the complex interactions between actors in processes of technoscientific innovation is that of the "triple helix" (Etzkowitz and Leydesdorff 2000). From this perspective, processes of scientific innovation are characterized by the intersection of three distinct spheres: the scientific, the one pertaining to the government and state and the one represented by industrial sectors and the market. Over the years, typical patterns of relationships among these three spheres have gradually become more and more varied, especially since the market actors and scientific institutions have seen radically transformed their practices and structures under the pressure of the commercial interests surrounding new technologies and patents. Universities dealing with business projects are the visible side of larger changes that have contributed in modifying the role of academic and scientific institutions and their relationships with the market. These transformations have not followed a linear trajectory but have been characterised by the emergence of contrasting relationships between the marketplace and science, generating a global picture with the contours of a 'post-academic' scientific research era (Ziman 2000; Bucchi 2010).

Nanofab represents an attempt to generate at local level a new organizational form of relationship between the three distinct spheres involved in innovation and described by Etzkowitz and Leydesdorff. In fact, from an organizational point of view, the Nanofab project started by recognizing that existing innovation models – i.e. those represented by science parks (Löfsten and Lindelöf 2002; Phan et al. 2005) – already present in Italy since the 1990, were not able to generate effective processes of technology transfer from academia to firms and industries. This issue has been explicitly addressed by one of the founders of Nanofab:

[Science parks] include many people who generally do not do research. These people have some connections with the world of research and they usually offer these connections to firms. Sometimes, among other things, these connections are not exhaustive, are not likely to represent a complete, or even simply adequate, supply of research as it is for academic universities. Because they have just few contacts, they tend to channel everything to those contacts, and thus the offer is not adequate to the firms' demand (Interview nr. 5).

While science parks generally do not hold specific scientific expertise and work often as intermediaries in the innovation process, Nanofab was different because it was designed with the aim of eliminating the step of intermediation in the exchange between academia and those actors dealing directly with the market. Seen from this point of view, the Nanofab trajectory has emerged as the result of the search for a different layout of relationships between institutions, academia and business, a new configuration eliminating those rigidities characterizing the usual science parks' activities. Above all, the main aim of Nanofab's founders was to be able to manage within the organization both scientific issues and managerial and business decisions in order to coordinate in the most efficient way their resources and also to build a more direct collaboration between scientific knowledge and market dynamics.

In the case of innovation patterns typical of science parks, these entities play the role of mediators and catalysts of research and market activities that are carried out by other agents. Science parks represent some sort of 'gatekeepers' in the innovation process (Herzog 1981), acting as filter for the possibility of cooperation between the different spheres and actors. Is easy to figure out one of the main limitations of this model, that consists in the lack of skills and scientific knowledge within the organization: mediators can only serve as facilitators for other actors and, generally, they have neither margins in the scientific choices nor a sufficient flexibility in managing relationships between firms and researchers.

The model of innovation and technology transfer that emerges from the actual experience of Nanofab offers a very different configuration of the spheres involved in the project. This configuration embodied by Nanofab can indeed be described as a 'trilateral network' (Etzkowitz and Leydesdorff 2000) in which the management of relationships between actors is ensured by the presence of a 'boundary organization' (Guston 1999, 2001; Tuunainen 2005), which is able to handle autonomously both scientific choices, institutional relationships and material resources, as well as marketing strategies and interactions with private firms. The configuration of relationships between the three spheres involved in the process of innovation is characterized by the presence of a "boundary organization" – Nanofab – which is crucial to help in maintaining active and constantly updated interactions between the different actors involved. These interactions are carried out thanks to an organization able to redefine and constantly renovate the boundaries between the different spheres involved (Guston 1999, 2001). As a 'boundary organization', Nanofab adopts an organizational approach which leads to a more effective and functional interaction between usually distinct spheres and which reallocates within its own organization tasks, interests and also responsibilities implied in the development of new technologies and products.

This new organizational configuration, which provides the possibility to develop a more direct interaction between political institutions, scientific actors and the market, has been designed with the primary aim of facilitating the involvement of private firms in the process of innovation and technology transfer. As we shall see later, one of the key aspects of Nanofab concerns precisely the development of organizational tools and practices aimed at enrolling firms and private businesses more and more in the process of entrepreneurial innovation. As a "boundary organization", Nanofab is able to redefine trust schemes, responsibilities, duties and interests among the actors belonging to the different spheres involved in innovation processes. Unlike other models of interaction, Nanofab's features include the ability to manage tasks and responsibilities within its own structure, without requiring the help of other mediators. In this sense, Nanofab's activities are carried out through what has been described as a work of 'hybrid management' (Miller 2001, 487). According to Miller, it is possible to analytically distinguish four distinct steps characterising more flexible and malleable forms of management in innovation processes: "hybridization", "deconstruction", "boundary work", and "cross-domain orchestration" between different spheres. These different operations are simultaneously present in the work of Nanofab, which in fact – as we will see empirically in the next section – plays a continuous effort of alignment and re-alignment of different cultures, practices and needs and, in so doing, it is crucial in defining and allocating responsibilities and trust between the different actors involved.

Now I will explain the role of Nanofab as a "boundary organization" and what this means for the emerging configuration of responsibility in the nanotechnology domain. In doing so, Nanofab's 'hybrid management' will be described in detail and the work developed at some specific stage of collaboration with firms and industries will be illustrated.

10.4   Configurations of Cooperation and Responsibility

To better understand the relationship between innovation, responsibility and organizational configurations in the case of nanotechnology at Nanofab, we can now focus on those specific steps through which Nanofab generate cooperation with private firms. One of the crucial aspects in the work of Nanofab consists in engaging firms from the private sector within the innovation process, establishing different possible forms of collaboration. Among these forms, the most direct and immediate is in that defined as the 'order'. In this case, Nanofab agrees with firms to adapt nanotechnology-based technical options to firm's products, processes or machineries and maybe this first step is just a simple exploratory analysis. These 'orders' consist therefore of secondary activities with usually a budget of between 5,000 and 30,000 euros. These collaborations frequently start with a meeting between Nanofab's representatives and the heads or managers of the companies expressing interest. Nanofab's representatives are often accompanied by experienced researchers in the field relevant for the collaboration and for the firms' sector, while business leaders are sometimes together with managers from the 'research and development' area. The first meeting is crucial in the success of their potential collaborations, because it sets the framework for the future developments, as Nanofab's Marketing Director has commented during our interview:

These initial meetings are always the 'key'. Because in these meeting you should always have the opportunity to make it clear that you know your job and to do it well enough and also that you're able to understand their needs, even in technical terms (Interview nr. 4).

Looking at these initial meetings, and especially at the very first meeting with firms, it is possible to recognize at least three crucial aspects in the development of collaborations. First, during these meetings Nanofab's staff have to figure out the needs or problems that firms could be interested in addressing and resolving. Second, it is necessary to understand which area of Nanofab's scientific knowledge could be useful to solve these problems. Finally, it is necessary to define a setting that can foster the relationship between Nanofab and the firms' representatives. This requires the parties involved to outline the boundaries of reciprocal responsibilities for collaboration. The head of Veneto Nanotech summarized these different aspects as follows:

Number one: offering clues that encourage companies to innovate using these technologies. Number two: collecting requests about technologies currently not available, because that too is a lever capable to raise funds and to enable other activities, which in any case will have some industrial interest. Number three: having good meetings, spotlighting opportunities that can only be achieved with our technology (Interview no. 1).

These different steps need to be constantly bound together. Thus, Nanofab's staff are required to perform continuously a work of 'alignment' between Nanofab's scientific potential and the needs and expectations of the business firms. Indeed, seen from the perspective of Nanofab's Marketing Director, the worst way to proceed is to talk about technical solutions that cannot realistically be achieved at the centre:

What you *cannot* do is to promise results to companies that afterwards Nanofab's researchers believe they can't achieve. So there must always be a very strong *alignment* between things that are promised and [the results that are] transferred to the companies (Interview no. 6).

The crucial importance of this continuous 'alignment' emerges repeatedly in the words of those involved in business meetings on behalf of Nanofab. Often people interviewed had stressed that this work of alignment does not just regard the technical adequacy. In the process of innovation, 'to align' means to identify not only those technical and scientific aspects that can offer a solid base for future collaboration, but also to define roles, responsibilities and competencies between the parties involved. As shown by the following observation from another Nanofab staffer, it is possible to find out at least another dimension characterizing the work of 'hybrid management' during the innovation process:

Well, let me say that in order to have everything running smoothly there should be an alignment between all actors, in the sense that you have to set the whole research framework. There must be clarity in relationships and roles, that is one should know exactly where it is going, how, with what resources, etc (Interview nr. 6).

From the point of view of one of the researchers who usually participates in meetings with firms, this process of alignment is a very complex part of the work. Indeed, one of the problems implied in this activity is the fact that companies often do not have a clear question to ask to Nanofab, but rather they have a number of problems that might not be resolved through the knowledge developed within Nanofab's laboratories:

> Sometimes [firms] do not have such clear ideas. The problem is that they often come with a list of problems, they come with a specific need and are asking if there is actually a solution to solve them. The step to find a solution for that need is often not easy. This means that if someone asks you to make an improvement that combines things that maybe we know are contradictory, it is pretty difficult [to find a solution]. Operationally, maybe there is no product, there is no material, there is nothing to match that need (Interview nr. 9).

This work of alignment includes, in addition to the need not to present objectives that cannot be technically achieved, even the necessity to clarify the responsibilities between different actors involved and, therefore, also to outline reciprocal risks with regard to the expected outcomes of the collaboration:

> Besides having gained the firm's confidence, you have to be able to make the company accept its responsibility. And this is a second important step. Because in the first one you should gain its confidence saying: 'we're ready'. After a while, when confidence has been gained, you must let them know that this is a research activity and that there is some risk implied and they have to be aware of this risk. Then it's clear that this topic is a bit nuanced at first, and it should gradually be made sharper once the relationship is established (Interview nr. 6).

During the meetings another very important dimension is the building of the credibility of the Nanofab team and the definition of the reciprocal responsibilities with respect to the risks entailed in the innovation process. The very first meeting, mainly participated by firms' managers and directors, commonly focuses on general and non-specific issues. Only during subsequent meetings can the collaboration begin to deal with more scientific details, allowing the parties to discuss more directly the specifics of technical problems, as one of the Nanofab researchers has outlined:

> There can be a general meeting in which the counterparts are directors or CEOs, or people of this kind that always know their own realities very well. Maybe they do not go into the specifics, but they are anyway able to sense the opportunities. Subsequent meetings generally include a technical content that is a bit higher, because this is the appropriate stage to go into the details of the matter (Interview no. 4).

However, also in the case of the initial meetings, where confidence and responsibilities need to be cleared up, scientific aspects are not absent, but can be mobilized indirectly in order to build the overall framework of collaboration. Scientific and technical details are rather pivotal tools around which Nanofab managers both generate confidence in their work and develop the configuration of reciprocal responsibilities in the innovation. One example is when, during the meetings, technical questions are introduced by firms to test the reliability of Nanofab. A Nanofab manager has observed:

> Then we try to increase our credibility, typically by exhibiting credentials and saying things like: 'we've already done this… we've also done that… and now we'll do this other thing, without naming brands of companies belonging to sectors where we are involved in. And then usually firms ask one or two questions with a very technical subject, and we should to be able to give a good answer, because it's at that exact moment that you gain the firm's confidence. When you have answered this first question well you have built up a lot of confidence (Interview no. 1).

As we can see the from these observations regarding how Nanofab manages a specific stage of its collaboration with firms, the establishment of confidence and trust as well as the definition of reciprocal responsibilities represent critical issues for building a successful path to innovation. Moreover, these considerations clarify the role of some of the critical aspects of technology transfer in nanotechnology and also the way the negotiation of responsibility is part of the work. More specifically, from the previous accounts we can see that the interaction with firms requires the simultaneous management of heterogeneous aspects, including social, technoscientific and commercial logics. During the meetings described above, the exchange fluctuates continuously from a technical level to a different one implying the definition of issues such as confidence, risk and responsibility, thus highlighting that responsibility is at stake throughout every stage of the transfer of nanotechnology from academia to the market.

10.5 Conclusions

The discussion about the setting up of Nanofab and about its activities offers the opportunity to develop at least two more general considerations regarding reciprocal influences and links between nanotechnology, organizational configurations and the emerging patterns of responsibility. First of all, as we have argued from the description of Nanofab's activities, the creation of a new organization able to develop novel patterns in innovation has required the definition of new patterns of relationships as well as of mutual responsibilities and expectations between the actors involved. In the Nanofab experience, the scientists' initial idea of creating this organization required the activation of a whole network of relationships to support the project. Together with the network, it has been necessary to foster a new geography of responsibilities among the actors involved and it required also the development of hybrid forms of innovation.

A second point that has emerged from the analysis regards the role of the configurations of responsibility in fostering the cooperation between actors belonging to different spheres and with different internal logics. By focussing on the description of the initial stages of the collaboration between Nanofab and private firms, we have seen that the establishment of a framework of mutual accountability between all the parties involved is one of the basic elements for the building of fruitful relationships. In particular we have also seen that, in the construction of the collaborative frameworks, the emerging configuration of responsibility develops hand in hand with the construction of trust and reciprocal recognition. This has been achieved both through the use of more traditional credentials and thanks to the mobilization of discursive strategies which were not necessarily based on the exchange of technical knowledge.

More in general, considering the Nanofab experience, we can recognize how responsibility in nanotechnological innovation represents a multiple and differentiated entity, which is relevant for different levels of the process of innovation. This reminds us of the multi-faceted and multi-layered nature of the concept of responsibility in innovation and, in particular, in nanotechnology.

Acknowledgements   This chapter is based on a research carried out in the period 2007–2009 and entitled: *A multidimensional approach to technological transfer*. The research has been carried out at the Department of Sociology, University of Padova in collaboration with the Free University of Bolzano and Trieste Synchrotron SCPA with the support the National Research Council – Institute for Physics (CNR-INFM) and the Ministry of Education, University and research (MIUR). I thank my colleagues who collaborated in the research and especially Paolo Volontè, who coordinated the project and national level and Federico Neresini, who was head of the research at the University of Padova.

References

Bijker, W., P. Volontè, and C. Grasseni. 2010. Technoscientific dialogues expertise, democracy and technological cultures. *Tecnoscienza. Italian Journal of Science and Technology Studies* 1(2): 121–140.

Bucchi, M. 2010. *Beyond technocracy: Science, politics and citizens*. London: Springer.

Doubleday, R. 2007. Organizing accountability: Co-production of technoscientific and social worlds in a nanoscience laboratory. *Area* 39(2): 166–175.

Etzkowitz, H., and L. Leydesdorff. 2000. The dynamics of innovation: From National Systems and 'Mode 2' to a Triple Helix of university – industry – government relations. *Research Policy* 29(2): 109–123.

Faunce, T.A. 2007. Nanotechnology in global medicine and human biosecurity: Private interests, policy dilemmas, and the calibration of public health law. *Journal of Law, Medicine, and Ethics* 35(4): 629–642.

Guston, D.H. 1999. Stabilizing the boundary between US politics and science: The role of the office of technology transfer as a boundary organization. *Social Studies of Science* 29(1): 87–112.

Guston, D.H. 2001. Boundary organizations in environmental policy and science: An introduction. *Science, Technology, and Human Values* 26(4): 399–408.

Herzog, A.J. 1981. The 'gatekeeper' hypothesis and the international transfer of scientific knowledge. *The Journal of Technology Transfer* 6(1): 57–72.

Knorr-Cetina, K. 1981. *The manufacture of knowledge: An essay on the constructivist and contextual nature of science*. Oxford: Pergamon.

Latour, B. 1987. *Science in action*. Cambridge: Harvard University Press.

Latour, B., and S. Woolgar. 1979. *Laboratory life. The social construction of scientific facts*. London: Sage.

Law, J. 1987. Technology and heterogeneous engineering: The case of Portuguese expansion. In *The social construction of technological systems. New directions in the sociology and history of technology*, ed. W.E. Bijker, T.P. Hughes, and T.J. Pinch. Cambridge, MA: MIT Press.

Law, J., and J. Hassard (eds.). 1999. *Actor network theory and after*. Oxford: Blackwell Publishers.

Löfsten, H., and P. Lindelöf. 2002. Science parks and the growth of new technology-based firms-academic-industry links, innovation and markets. *Research Policy* 31(6): 859–876.

MacNaghten, P., M.B. Kearnes, and B. Wynne. 2005. Nanotechnology, governance, and public deliberation: What role for the social sciences? *Science Communication* 12(27): 268–291.

McCarthy, E., and C. Kelty. 2010. Responsibility and nanotechnology. *Social Studies of Science* 40(3): 405–432.

Miceli, V. 2010. *Distretti tecnologici e sistemi regionali di innovazione: il caso italiano*. Bologna: Il Mulino.

Miller, C. 2001. Hybrid management: Boundary organizations, science policy, and environmental governance in the climate regime. *Science, Technology, and Human Values* 26(4): 478–500.

Neresini, F. 2006. Starting off on the wrong foot: The public perceptions of nanotechnologies and the deficit model. *Nanotechnology Perceptions* 2(2): 189–195.

Nerlich, B. 2008. Powered by imagination: Nanobots at the science photo library. *Science as Culture* 17(3): 269–292.

Phan, P.H., D.S. Siegel, and M. Wright. 2005. Science parks and incubators: Observations, synthesis and future research. *Journal of Business Venturing* 20(2): 165–182.

Pinch, T.J., and W. Bijker. 1984. The social construction of facts and artifacts: Or how the sociology of science and the sociology of technology might benefit each other. *Social Studies of Science* 14(3): 339–441.

Rampersad, G., P. Questerb, and I. Troshania. 2010. Managing innovation networks: Exploratory evidence from ICT, biotechnology and nanotechnology networks. *Industrial Marketing Management* 39(5): 793–805.

Robinson, D.K.R. 2009. Co-evolutionary scenarios: An application to prospecting futures of the responsible development of nanotechnology. *Technological Forecasting and Social Change* 76(9): 1222–1239.

Roco, M.C. 2005. Environmentally responsible development of nanotechnology. *Environmental Science and Technology* 39(5): 106–112.

Roco, M.C., and W.S. Baingridge (eds.). 2001. *Societal implications of nanoscience and nanotechnology*. New York: Springer.

Selin, C. 2007. Expectations and the emergence of nanotechnology. *Science, Technology, and Human Values* 32(2): 196–220.

Tuunainen, J. 2005. Contesting a hybrid firm at a traditional university. *Social Studies of Science* 35(2): 173–210.

Ziman, J.M. 2000. *Real science: What it is, and what it means*. Cambridge: Cambridge University Press.

Chapter 11
Who Is Responsible? Nanotechnology and Responsibility in the Italian Daily Press

Simone Arnaldi

11.1 Introduction

This chapter discusses how the Italian daily press define the notion of responsibility in the news stories about nanotechnology and what the view of science-society relations that underlies such definitions is.

The starting point of this discussion is the acknowledgement that responsibility is ubiquitous in nanotechnology discourse. Despite its pervasiveness, this notion appears relatively undeveloped from two different points of view. On the one hand, responsibility is often presented as a generic concept, which is often defined indirectly by resorting to connected, but distinct, notions (e.g. ethical implications, health and environmental risks, precaution, sustainability, etc.). On the other, there is still limited empirical work on how responsibility is constructed in the different arenas where nanotechnology is debated (the existing research focuses on research organizations). Moving from these premises, the chapter tentatively links theoretical reflection and empirical research (i) by referring to the existing philosophical investigation on responsibility in the attempt to partially disambiguate the notion; (ii) by using this theoretical framework to guide empirical research into an exploration of the representations of responsibility in the media arena, thus extending the current focus of attention on the sites where responsibility in (nano)technology innovation is constructed.

The first section will briefly discuss the concept of responsibility through the lenses of some major works in philosophy and, to a lesser extent, sociology, emphasizing the challenge to responsibility in a technological age. The second section of the chapter will introduce some examples to illustrate the meanings attached to the

S. Arnaldi (✉)
Centre for Environmental, Ethical, Legal and Social Decisions on Emerging Technologies (CIGA), University of Padova, Viale Porta Adige 45, Rovigo, Italy

Istituto Jacques Maritain, Trieste, Italy
e-mail: simone.arnaldi@unipd.it

S. Arnaldi et al. (eds.), *Responsibility in Nanotechnology Development*, The International Library of Ethics, Law and Technology 13, DOI 10.1007/978-94-017-9103-8_11,
© Springer Science+Business Media Dordrecht 2014

concept of 'responsible nanotechnology', both in research and policy documents. In the same section, the more recent work on the construction of responsibility in local research and industrial practices will be presented. The third section aims at disambiguating the different meanings of responsibility as a preliminary work to better frame the relationship between technology and responsibility. It also introduces the object of the empirical investigation, that is news stories about nanotechnology in the Italian daily press and presents the results of our analysis, showing the patterns of the social division of responsibility-related labour and the role technology is assigned in this framework.

11.2   Responsibility and Nanotechnology

The search for a comprehensive overview of the concept of responsibility in history and across disciplines is far beyond the goals of this chapter (see e.g., Miano 2009 for a short history of the notion and a more detailed overview of the philosophical debate on responsibility in the twentieth century). The following paragraphs have a more modest, instrumental goal to find a working definition able to guide our empirical work of detecting the representations of responsibility in Italian news stories about nanotechnology.

Discussing how sociology has been dealing with responsibility, Strydom (1999) notes that the traditional definition of this notion refers to individual responsibility within informal and pre-institutional (e.g. friendship, family, kinship) or institutional contexts (e.g. occupational role). In the latter case, 'we are still dealing with a conventional form of individual responsibility insofar as it remains within the normative confines of a given institutional framework' (Strydom 1999, p. 68). Classic notions like Durkheim's functional division of social labour or Talcott Parsons's view of responsibility as a complex of duties associated to social roles are in line with this traditional view of responsibility. From a non-functionalist point of view, Weber's notion of 'ethics of responsibility' is squarely placed in an individual perspective, and focuses on the agent's calculative capacity of appraising means and ends, actions and their consequences. According to Strydom (1999, pp. 68–69), a post-traditional, but still individual, notion of responsibility, regards individuals who possess 'special knowledge, abilities, judgement, power or influence in particular domains of social life, [rather than] observing, traditional or conventional limits [...] take the initiative to shift the boundaries by assuming individual responsibility for the (re)design and (re)organization of institutions and social systems themselves with a view to the constant monitoring and the reduction or avoidance of negative features and effects'. For Strydom, public intellectuals or prominent individuals challenging established conventions exemplify this category of responsibility.

While Authors differ in wording, there is agreement that this individual idea of responsibility is radically and decisively challenged by the advent of 'modern science-based' technology. The changing nature of responsibility depends on the new and unprecedented features of a technology-infused world: 'socially, it is universal in the sense that it confronts all people with the same problem; spatially, it is planetary or global in the sense that is spans the whole earthly biosphere; and temporally, it is irreversible in the sense that it reaches into the future and potentially affects all future generations of humans, animals and plants' (Strydom 1999, p. 66).

Two of the most essential and well-known contributions to examine this original situation are probably those by Hans Jonas and Karl-Otto Apel. Apel stresses the interdependence of the 'applicability of modern technology to the social dimension' and 'the simultaneous process of the so-called *differentiation of social life into functional-structural social systems or, respectively, sub-systems*' (Apel 1993, p. 10), for 'individual actors in a sense cannot really be held *accountable* for these actions and activities in such a way as individuals have been held responsible for their actions according to traditional morals' (Apel 1993, p. 14). Jonas stressed famously the failure of two pillars of traditional responsibility: the capacity of the agent to foresee the consequences of its acting to some extent and to control the impact of its own acting, while the third pillar ('causal power') expands dramatically (Jonas 1984, p. 90). Jonas has connected this failure to the specific and original features of technology development and its consequences: collectivity, as technology development is typically a collective endeavour; indirect causation, as technology's effects characteristically occur as an outcome of long causal chains; uncertainty, as the type, scale, and probabilities of possible consequences of technology are unknown.

Arguments are largely convergent, although outcomes are not and the notion of collective (Jonas) and co-responsibility (Apel) have been considered contrasting by the Authors themselves (at least by one Author, see Strydom 1999). However, it does not seem a too long a stretch to say that their positions stress different critical aspects of the traditional view of responsibility in the technological age. Jonas stresses the epistemic dimension of the problem, i.e. the unmanageable uncertainty surrounding 'the extended scope of human intermingling with nature' (Pellizzoni 2010, p. 467). Apel stresses the systemic inability to ascribe responsibility and cope with the consequences of collective acting. However, both Authors sanction the modern association of responsibility and (technological) risk.

The transformative power that is attributed to nanotechnology makes this emerging field a perfect candidate to exemplify the far-reaching, collective, and uncertain technological venture that Apel and Jonas consider in their theory of responsibility. The awareness of the magnitude of nanotechnology's capacities is the reason why responsibility is everywhere in nanotechnology discourse and the imperative that nanotechnology should be managed responsibly pervades scientific literature and policy documents.

The Code of Conduct for responsible nanoscience and nanotechnologies research promoted by the European Commission (EC 2008) is a prominent example of such attention for a responsible nanotechnology. The Code states that '[n]anotechnology must be developed in a safe and responsible manner' (EC 2008, p. 3) and sets responsible development as a indefeasible goal to seize the opportunities raised by nanoscience and nanotechnology. The Code 'that N&N [nanoscience and nanotechnology] research is undertaken in the Community in a safe, ethical and effective framework, supporting sustainable economic, social and environmental development' (section 1). The Code does not provide a definition of responsibility, rather it establishes a connection between responsibility and a set of principles that should orient nanotechnology research: sustainability, precaution, inclusiveness, accountability. Acknowledging the uncertainty that surrounds nanotechnology impacts and drawing on these principles, the Code calls for 'a general culture of responsibility (section 4.1) that should be created in view of challenges and opportunities that may be raised in the future and that we cannot at present foresee' (EC 2008). The Royal Society, Insight Investment and the Nanotechnology Industries Association (NIA)'s Responsible Nanocode does not discuss what responsibility is, but, in a way that is similar to the EC Code of Conduct, which establishes domains and principles that are aimed to steer stakeholders (i.e. industry, in this case) so as to ensure a responsible development of products incorporating nanomaterials. Such principles detail the conditions for attributing responsibility to NIA member organizations and to instances that are internal to these organizations (accountability, transparency and disclosure), and for the responsible management of nanotechnology by member organizations (engaging with business partners, stakeholder involvement), as well as the domains which require responsible behaviour (worker health and safety, public health, safety and environmental risks, wider social, ethical environmental and health impacts). These policy documents are exemplary in that responsibility is not defined *per se*, but domains and principles of conduct are listed so as to indirectly define the boundaries and features of responsible practices in R&D, industrial production, and commercialization.

The reference to impacts and their governance is essential to this indirect definition. The debate on responsibility in nanotechnology development is indeed squarely located in the broader discussion on ethical, legal and social (ELS) and/or environmental, health, and safety issues (EHS) of nanotechnology. This is true for the European examples we mentioned above, and for the US experience alike. For instance, according to the Triennial Review of the National Nanotechnology Initiative, '[r]esponsible development of nanotechnology can be characterized as the balancing of efforts to maximize the technology's positive contributions and minimize its negative consequences. [...] It implies a commitment to develop and use technology to help meet the most pressing human and societal needs, while making every reasonable effort to anticipate and mitigate adverse implications or unintended consequences' (Committee to Review the National Nanotechnology Initiative 2006, p. 73). The following paragraphs illustrate two examples, among countless literature references from Europe and the US, dating back respectively to the initial and the more recent phases of nanotechnology as an object of public debate. The Roco and Bainbridge edited volume on the societal implications of nanoscience and nanotechnology (2001) is indeed indicative of the different characterisations of responsibility in the 'hybrid' policy-academic environment created by the NSF-sponsored reports and events that popped up after the National Nanotechnology Initiative was signed into law in 2000. Contributions to the report discuss 'social responsibility' in different terms, as a response to societal concerns which is part of scientists' and engineers' mandates (Mizrach in Tolles 2001, p. 181), including 'the presence of a responsible and respected scientific body with thoughtful statements about the reality of the options and possibilities arising from research in this area' (Tolles 2001, p. 188), as the incorporation of ethical principles and issues in professional societies guidance and in a curriculum for training 'new nanoscientists, nanotechnologists, and nanofabrication technicians' (Roco and Bainbridge 2001, p. 13), as collective, 'growing social readiness' for coping with 'the possible impact on business, society and the economy' (Canton 2001, p. 97), as public inclination to 'take a greater responsibility for the science and technology underway (taking what is often referred to as 'the public understanding of science' a step further)' (Glimell 2001, p. 159), and, finally, as a need to change the notion of legal liability because 'nano-mechanisms will test established assumptions about responsibility and control' (Suchman 2001, p. 215).

More recently, a report by the ObservatoryNano project straightforwardly links responsibility and ethical, social and legal implications of nanotechnology: '[t]aking on or placing responsibilities, then, are ways of making explicit ethical and social implications of science and technology in professional settings and on a broad societal level' (Malsch and Hvidtfelt Nielsen 2009, p. 4). In a coherent way, the report engages in two different lines of argumentation. On the one hand, this work discusses the limits of a traditional idea of responsibility as a matter of autonomous individuals in simple, well-defined social situations. On the other, the report identifies 'apparent responsibility issues', listing privacy, dual use, the security-freedom balance, intellectual property issues, distributive justice and environmental impacts as major concepts around which the notion of responsible development is organised. Who can be and who is held responsible for the nano-scientific and technological development? How can each stakeholder group take its responsibility? (Malsch and Hvidtfelt Nielsen 2009).

Although these documents, and many more, assert the compelling conjunction between technology and its social implications as the source of the responsibility discourse, it has been noticed that '[i]n spite of its growing centrality, the notion of responsibility is often left unproblematised and undefined' (Kjølberg and Strand 2011, p. 108). A recent, emerging stream of the literature attempted to close this gap and it has been examining how responsibility is articulated in research practices. Using ethnographic observation (McCarthy and Kelty 2010), and interviews either with nanoscientists (Kjølberg and Strand 2011) or with nanotechnology stakeholders (Foley et al. 2012), these scholars have attempted to understand how scientists conceive responsibility and how they translate this notion in practice.

Foley et al. (2012) categorizes responsibility around two dimensions related to the negotiable non-negotiable values, or tacit beliefs in research practice, i.e. related to the underlying understanding of the field, that are associated with nanotechnology research and to the different levels of society involved in implementing responsibility (individual, professional societies, macro-level collectives). By interviewing 45 organizations engaged in nanotechnology development in Maricopa County (USA), the Authors find a majority of statements related to responsibility pertaining to non-negotiable values and individual and professional societal levels, while 'actors in this community express macro-level or tacit beliefs embodied within the nanotechnology innovation process with limited frequency' (Foley et al. 2012).

Kjølberg and Strand (2011) attempt to dwell on how responsibility is perceived by nanoresearchers and outline three, empirically grounded 'broad notions of Responsible Nanoresearch'. The first notion frames responsibility in the terms of the 'traditional social contract of science': society benefits from scientific knowledge, so 'the most responsible thing to do is to change as little as possible' and for what concerns risks, 'the reasonable amount of resources has been allocated to risk-research; the expected benefits exceeded the expected risk and so forth' (Kjølberg and Strand 2011, p. 109). The second notion considers responsibility in terms of 'deliberation across levels and sectors'. According to this definition, 'if one managed to institutionalise deliberation between sectors and levels, nanoresearch as a profession would be responsible'. From this point of view, 'responsible nanoresearch is taking seriously other actors as conversational partners and co-producers of the nanofuture, founded on an understanding that nanoresearch is one among many relevant fields of knowledge, and the scientific one among many possible framings of the development of nanoST' (Kjølberg and Strand 2011, p. 110). The third and last notion of responsibility is rooted in scientists' 'increased awareness of moral choices' and 'starts from the acknowledgement of how this shortcoming ultimately leads to personal moral judgement and choice', thus suggesting 'the necessity of an internally initiated and personal thinking that goes beyond action' (Kjølberg and Strand 2011, p. 110).

McCarthy and Kelty (2010) explored the co-production of individual and collective responsibility by the nanotechnology researchers who founded the Center for Biological and Environmental Nanotechnology (CBEN) and, at a later stage, contributed to the establishment of the International Council on Nanotechnology (ICON). Their research explored what 'do-able' responsibility (McCarthy and Kelty 2010, p. 409) is for scientists and how this notion draws them to 'two separate but entangled ideas: the risks that nanomaterials pose to biology and the environment [the research topic at CBEN], and the risks that research on this area poses to the health of nanotechnology itself' (McCarthy and Kelty 2010, p. 409), and how this double-edged, individual responsibility is built into the institutional settings of CBEN and ICON.

11.3   Exploring the Dimensions of Responsibility in the News

The chapter examines the representations of nanotechnology in the Italian daily press as its object of reference and attempts to respond to two distinct research questions, thus extending to the media discourse the existing research work on the configurations and meanings of responsibility in nanotechnology research practice.

However, such an effort is faced with a challenge: almost no news story deals explicitly and primarily with responsible research and innovation as such. They are rather descriptions of scientific activities, research results, impacts of innovations, and policy choices. Therefore, any research approach has to elicit the meanings of responsibility in the media coverage by observing how nanotechnology research and policy is described and, subsequently, by examining how these characteristics refer to the existing classifications of the notion of responsibility. This analysis is here intended to answer two research questions. The first one concerns the salience of the different types of responsibility as they are represented in the news coverage. The second one regards the logic in the relations between science, technology, and society that underlies the representations of responsibility that appear in the news.

Having regard to this twofold goal, the chapter follows the Kjølberg and Strand typology (Kjølberg and Strand 2011) to identify what type of responsibility the media coverage associates with the work in the nanotechnology field. Drawing on their descriptions of the three 'broad notions' of responsible nanoresearch, the Author identified, three groups of codes to map statements made by social actors pictured in the news and assign them to one of the three types ('traditional contract of science', 'deliberation across levels and sectors', 'increased awareness of moral choices'). Opinion pieces were considered as one single statement. Single statements in interviews were coded separately when they had different content (e.g. if two contiguous sections or questions dealt with the same content, they were treated as a single statement).

The codes referring to responsible nanoresearch as 'traditional contract of science' concerned the following issues: science and technology as leading to social benefits, research as progress of scientific knowledge, research practice and collaboration as a way to deliver science's promised benefits. Also, praise of social actors (public administration, business) for supporting science, as well as blame of social actors for not supporting/creating obstacles to science are included in this category. Risk research as a way to deal with potential risk was also included here. The codes referring to responsible research as 'deliberation across levels and sectors' concerned issues related to radical uncertainty of nanotechnology impacts, as well as 'societal issues' at large, like communication, public dialogue, public perception. Claims to adopt regulatory measures, as well as references to specific ethical and social issues (e.g. distributive justice) were included here. Statements were related to responsible research 'as increased awareness of moral choice' when they mentioned explicitly moral choices or responsible management as a necessary condition for nanotechnology research and development.

The sample of news stories to be analysed, both routine news reports, op-ed, in-depth interviews or feature articles, were retrieved from a broader sample of news stories about nanotechnology in three major Italian daily newspapers (*Corriere della Sera*, *Il Sole 24 Ore*, *La Stampa*) and two Italian newswires (*ANSA* and *ADN Kronos*). Both the newspapers and the newswires are indexed in the database Factiva and a total of $N=218$ nanotechnology stories was retrieved by using a boolean search string based on the work of Dudo et al. (2011), which has the merit of reducing the number of 'false positives' and the work needed to manually screen each story retrieved by the boolean term search to remove them (see Appendix A). The search was run for news stories from 1 January 2001 to 31 March 2012. This chapter analyses a subsample of $N=116$ news stories from 1 January 2008 to 31 March 2012.

The following sections discuss separately statements by nanoscientists and from other social actors.

11.4 Scientists Speak Out on Responsibility

A total of N = 128 statements by scientists were coded. Statements of Italian scientists working in their home country or abroad approximately amount to the 83 % of the total. The remaining statements were made by foreign nanoscientists (Table 11.1).

With no doubt, nanoscientists frame their work overwhelmingly in the terms of 'traditional contract of science', with 86 % of all statements that can be referred to this view of responsibility. Accordingly, scientists are primarily responsible to deliver science and technology benefits, through overcoming the difficulties and uncertainties surrounding this emerging technological field. Not surprisingly, reference to expected benefits of nanotechnology is the greatest single issue that is referred to in scientists' statements (49 %). On the one hand, nanotechnology dramatically improves our technological means, e.g. in the health field.[1]

> Nanotechnologies have the potential to deliver drugs exactly where they are needed, but also to detect early signals of cancer in good time as compared to traditional diagnostic tools (Ansa 2011a).

Table 11.1 Statements by nanoscientists in the news and associated notions of responsibility

Category of responsibility/groups of codes	No. of statements per nationality of source		
	Italian	Foreign	Total
Traditional mandate of science	**93**	**17**	**110**
Science as a curiosity-driven enterprise	9	2	11
Science as a driver to social benefits	54	9	63
Further research suffices to cope with risks	3	4	7
Social actors are ineffective to support science	5	0	5
Social actors are effective to support science	8	0	8
Good research practice as a condition to deliver science benefits	5	0	5
Communication is needed to disseminate progress and avoid rejection	2	2	4
Direct dialogue with citizens favours the public understanding of science	3	0	3
Public perception is important as it determines acceptance/rejection of nanotechnology	4	0	4
Public deliberation across levels	**11**	**5**	**16**
Uncertainty surrounds nanotechnology and its impacts and cannot be solved	11	5	16
Individual moral choices	**2**	**0**	**2**
R&D in emerging areas of science requires early ethical analysis	1	0	1
Innovation should be managed carefully to avoid adverse consequences	1	0	1

---

[1] All English translations are made by the Author of this chapter.

By reducing the dimensions of components through the use of nanostructured materials, she [Judit Puskas, Author's note] explains, 'we can develop materials with unprecedented properties, thus solving the typical complications related to the use of silicone' (Ansa 2011b).

Besides expected benefits stemming from future scientific breakthroughs, current impacts on economic growth of regions and companies involved in nanotechnology research are referred to prominently in the news stories. Interestingly enough, scientists themselves see this type of return as a essential part of their job.

Thanks to this invention – Claudio Cagliero says –, we hope to create some job opportunities in Italy (Favro 2008a).

Siena Nanotech is our contribution to this goal: it will foster development, it will create new, high-skilled jobs, and wealth for the region (Bianchi Roma 2009).

In four years we have attracted more than 100 million [dollars] of funding to Texas, I have more than 100 researchers in the lab and we have started three firms, with an economic impact creating from 4 to 6 thousand jobs, according to official figures of the Texas State Government. Therefore, we think we have had some modest accomplishments (Grossano 2010).

While there are no significant differences between Italian and foreign scientists in stressing that their responsibility is first and foremost to make good research and deliver the benefits of science, only Italian scientists assess other social actors' responsibility in creating the conditions for them to successfully accomplish their tasks (10 % of the total statements). Such evaluations include both praise for alleged positive performances and blame for negative ones.

Today, there is Airc [Italian Association for Cancer Research, a charity funding medical research, Author's note]. Otherwise, oncology research would be impossible in Italy (Rizzato 2011).

The Regional Government of Liguria does not invest on nanotechnology and thus the region misses an important opportunity for development. Other regional governments behave very much differently (Bompani 2009).

This region – Maiani states [the President of the National Research Council, Author's note] – is a leader in research support (Ricci 2010).

The 'traditional contract of science', and the place it assigns to scientific knowledge, also frames scientists' views of risk assessment and public engagement. From this perspective, statements concerning risk issues assume that risks can be controlled by allocating more efforts to risk-research. Also, public engagement and social perception are defined in the deficit model's terms, emphasizing one-way outreach and communication. When communication (rarely) regards risk issues, the public can acknowledge that the scientific community is concerned with such issues, but no institutional change in the way science is governed is mentioned: the goal is 'defensive', to avoid public rejection of this emerging technology.

[It is] a free exhibition open to everybody, which will aim at bringing nanoscience and technology marvels to the public (Favro 2008b).

It is an unprecedented preventive action for a new technology: that's why governments, companies, and researchers are studying how to maximise benefits and minimize harms of emerging nanotechnologies (Paterlini 2011).

> We do not want to share the fate of GMOs, which were rejected, albeit irrationally, by the public (Jacchia 2008).

When uncertainty surrounding nanotechnology and its impacts is concerned, a minority of statements acknowledge its unattainable nature. However, this awareness does not apparently lead to an open recognition of the need to change the institutional framework of nanotechnology governance, as specified in the notion of responsibility as deliberation across levels and arenas.

> I hear most often announcements like 'a new generation of much faster computers will be based on this innovation', 'some day a new therapy for cancer will be produced stemming from these results'. However, it is always an indeterminate future (Grassia 2008).

> We truly ignore almost completely the life cycle of nanoparticles, their destination, and their effects after their use, e.g. as a consequence of their accumulation in the body (Caprara 2008).

> In the present conditions, an unambiguous definition of nanomaterial for regulation is difficult if not completely impossible (Milani 2011).

The analysis shows therefore that it is extremely rare that scientists depart from the traditional view of science as a domain separate from society, and that freedom and resources should be given to science in order to bring its benefits to society and, also when the capacity of science to reduce uncertainty is questioned, this does not lead to a plea for modifying the current, expert-centred, model of technology governance. Therefore, e.g. when discussing uncertainty, there is a certain degree of ambiguity in scientists' take on issues related to nanotechnology development, which cannot be straightforwardly linked to an innovative view of the relationships between science and society.

We were able to find only two statements that could exemplify the third broad notion of responsible nanoresearch as defined by Kjølberg and Strand (2011), in which they call responsibility an 'increased awareness of moral choices'. The characteristic of this notion is its subjective framing: responsible research is a matter of individual choices, besides (and beyond) social roles and institutional settings. Accordingly, we coded only statements that were explicit either in mentioning ethical assessment for science and technology or in acknowledging that every innovation has to be responsibly managed. As we mentioned above, only two out of 128 statements were included in this group.

> [O]ne should remember how much these territories are treacherous and how much ethical assessment is fundamental as well (Beccaria 2008).

> It is evident that any human innovation has to be managed appropriately so that the advantages that it creates are not transformed into damage (Caprara 2008).

11.5   Other Voices on Responsibility

A minority of statements (N=29) were made by other social actors, representing business (N=18), policy-makers (N=6), citizens and civil society organizations (N=5). Apart from business representatives, whose statements were all focused on the nanotechnology opportunities and on the need to support nanotechnology development, this minority of actors does not voice the 'traditional contract of science', but rather calls for public engagement and precaution. Interestingly enough, the radical positions are the ones which are covered in the press.

'For years, members of the [European, Author's note] Parliament have requested a fair regulation: it is time for the European Commission to listen to the Parliament and the citizens on this issue', Kartika Liotard, the proposer of the bill and member of the European United Left group, said (Ansa 2010).

For all these reasons, food authorities must stop the use of nanoparticles that are not sufficiently tested in the food industry (Jac 2008).

Ordinary citizens are left alone to ask for shared decision-makers on the decision to establish a facility for nanotechnology research in agriculture and to allocate the potential benefits of such an initiative.

Everything can be done, but you have to involve farmers and not to take top-down decisions. Let's see, and hope that this science park and nanotechnology stories do not benefit only the landowners (Cremonesi 2012).

Also in this case, there is only one news article (an op-ed) which can be associated with a view of responsible nanoresearch as a moral choice. The opinion piece is, remarkably, by a theologian who comments on the results of the well-known Scheufele et al. (2009) research on the impact of religiosity on attitudes towards nanotechnology:

Technology is a tool and it cannot become the master. In the uncharted sea of technology applications in we are poised to sail, the beacon to head for is individual freedom, of this tangible individual, not of another, hypothetical one, which might be nicer but different from the original self. Personal freedom is the most precious thing we have and we must not sell it to anybody, not even technology (Mancuso 2008).

11.6   Business as Usual and Divided Voices: Closing Remarks

While different categorizations, sampling and coding strategies can highlight different features of the media coverage, three broad aspects can be stressed.

Firstly, the 'traditional contract of science', in Kjølberg and Strand's terms, is overwhelmingly dominant in the coverage. Responsibility is outlined in terms of the scientists' professional role and that the models and tasks associated with such a role are defined by a social mandate that reflects a technocratic view of the relationships between science, technology and society and supports a division of scientific and moral labour assuming that scientists are primarily responsible to deliver science benefits and that scientists are the ones most qualified to make decisions about priorities within research. As a consequence, scientists are entitled to outline the boundaries of other social actors' responsibility, which is to support science and technology development, and to blame them if they fail. Emphasis on public education and communication can be seen as an attempt to construct the public's role in terms of not being irrational and acknowledge the inherent worth of scientists' work.

Secondly, this division of labour also limits the number and variety of topics on which different social actors can be rightfully considered as sources for the coverage. More specifically, the discussion of radical uncertainties surrounding the nanotechnology enterprises, of precautionary measures, of new institutional arrangements for deliberation on science and technology, is left entirely to civil society organizations, citizens, and humanities scholars.

Thirdly, if we consider Kjølberg and Strand's notions of responsible nanoresearch, then we see how the discussion almost exclusively concerns institutional configurations, irrespective of the traditional division of technoscientific labour or the institutionalisation of deliberation and communication between levels and sectors in nanotechnology policy and research. What is missing almost completely is what Kjølberg and Strand define as 'the acknowledgement of how this shortcoming ultimately leads to personal moral judgement and choice' (Kjølberg and Strand 2011, p. 111) and the awareness that this dimension is the ground in which responsible discourse and action are planted.

Appendices

Appendix A: Boolean Search Term Used to Gather Articles from Factiva Database

atleast3 nanotec* OR atleast3 nanosci* AND ((nanoscal* OR nanocristal* OR nanotub* OR nanomat* OR (nanometr*) NOT (nanometr*/N15/luce or nanometr*/N15/ laser or nanometr*/N15/lunghezza or nanometr*/N15/UV or nanometr*/N15/onda) OR nanodot* OR nanomed* OR nanopart* OR nanofil* OR nanoing* OR nanocomp* OR nanoelettr* OR nanobot* OR nanorobot* OR nanomacch* OR fulleren* OR buckminsterfulleren* OR fullerit* OR carta/N2/grafen* OR grafen* OR buckytube* OR assembla*/N2/molecolar* OR fabbrica*/N2/molecolar* OR micromacch* OR quantic*/N2/dot* OR fil*/N2/quantic* OR pozz*/N2/quantic* OR sub micron OR (atom* adj5 manipol* OR atom* adj5 muov* OR atom* adj5 moss* OR atom* adj5 fabbric*) OR (microscopi* adj3 atomic*) OR (microscopi* adj3 tunnel) NOT (bomb/N10/atomic* or arm*/N10/atomic* or central*/N10/atomic* or bomb*/N10/ nuclear or arm*/N10/nuclear nuclear or central/N10/nuclear or nanosecond* or apple or ipod or mp3)) or digest or notizi*/N2/brev* or brev*/N2/economia or brev*/ N2/finanz* or brev/N2/mercat* or mostre or appuntamenti or rassegn*)

Appendix B: Searched Outlets

Adnkronos – General News (Italian Language) Or Publication Adnkronos – Health News (Italian Language) Or Publication Adnkronos – Labor News (Italian Language) Or Publication Adnkronos – Sustainability (Italian Language) Or

Publication AKI – Adnkronos International (Italian Language) Or Publication ANSA – Agricultural News Service (Italian Language) Or Publication ANSA – Economic and Financial Service (Italian Language) Or Publication ANSA – Entertainment News Service (Italian Language) Or Publication ANSA – Foreign Affairs News Service (Italian Language) Or Publication ANSA – Health Service (Italian Language) Or Publication ANSA – PMI News Service (Italian Language) Or Publication ANSA – Political and Economic News Service (Italian Language) Or Publication ANSA – Politics News Service (Italian Language) Or Publication ANSA – Regional Service (Italian Language) Or Publication ANSA – Sports News Service (Italian Language) Or Publication Corriere della Sera (Italian Language) Or Publication La Repubblica (Italian Language) Or Publication La Stampa (Italian Language)

References

Ansa. 2010. Ue: Parlamento, no pecora Dolly e progenie in piatto europei. *Ansa – Agricultural News Service*, July 7.

Ansa. 2011a. Tumori: parte progetto su nanotecnologie per diagnosi precoce. *Ansa – Health Service*, December 21.

Ansa. 2011b. Tumori: seno, nanotecnologie per protesi ricostruzione. *Ansa – Health Service*, October 5.

Apel, K.-O. 1993. How to ground a universalistic ethics of co-responsibility for the effects of collective actions and activities. *Philosophica* 52(2): 9–29.

Beccaria, G. 2008. Sfida all'ultimo limite È la nostra fantasia Cingolani, direttore dell'ITT: non ci limiteremo a copiare la natura La prossima fase è superare le architetture dell'evoluzione stessa. *La Stampa*, February 27.

Bianchi Roma, E. 2009. A Siena la rampa di lancio delle nanotecnologie. *La Repubblica*, January 26.

Bompani, M. 2009. Nanotecnologie, il futuro è piccolissimo – Genova crocevia di 300 scienziati: 'Ma la Regione non finanzia questo settore'. *La Repubblica*, July 28.

Canton, J. 2001. The strategic impact of nanotechnology on the future of business and economics. In *Societal implications of nanoscience and nanotechnology*, ed. M. Roco and W.S. Bainbridge, 91–96. Arlington: National Science Foundation.

Caprara, G. 2008. Se la fede frena le nanotecnologie. *Corriere della Sera*, December 9.

Committee to Review the National Nanotechnology Initiative. 2006. *A matter of size: Triennial review of the national nanotechnology initiative*. Washington, DC: The National Academies.

Cremonesi, F. 2012. Castel Cerreto dal 'Batistì' di Olmi alle nanotecnologie. *Corriere della Sera*, March 20.

Dudo, A., S. Dunwoody, and D.A. Scheufele. 2011. The emergence of nano news: Tracking thematic trends and changes in US newspaper coverage of nanotechnology. *Journalism and Mass Communication Quarterly* 88(1): 55–75.

European Commission. 2008. *Commission recommendation on a code of conduct for responsible nanosciences and nanotechnologies research*. C(2008) 424 final.

Favro, G. 2008a. La camicia pulita per sempre Tessuti "anti-tintorià' e nuovi medicinali: la rivoluzione delle nanotecnologie I progetti all'avanguardia in Italia. *La Stampa*, June 19.

Favro, G. 2008b. L'arte gioca con l'invisibile Tecnologia. Il nanotech inventa un filone estetico nell'universo dei miliardesimi di metro. *La Stampa*, February 6.

Foley, Rider W., Ira Bennett, and Jameson M. Wetmore. 2012. Practitioners' views on responsibility: Applying nanoethics. *NanoEthics* 6: 231–241.

Glimell, H. 2001. Dynamics of the emerging field of nanoscience. In *Societal implications of nanoscience and nanotechnology*, ed. M. Roco and W.S. Bainbridge, 156–160. Arlington: National Science Foundation.

Grassia, L. 2008. Questi i problemi ancora da superare. Cerrina, scienziato e businessman: 'La realtà a volte è più contraddittoria'. *La Stampa*, February 27.

Grossano, L. 2010. Ferrari, cervello in fuga, ho fondato impero nanotech. *Ansa*, November 8.

Jac, A. 2008. Cappuccino con la schiuma perfetta, ecco il 'nanofood'. *Corriere della Sera*, June 26.

Jacchia, A. 2008. Nanotech, il grande affare. *Corriere della Sera*, June 26.

Jonas, H. 1984. *The imperative of responsibility: In search of an ethics for the technological age.* Chicago: The University of Chicago Press.

Kjølberg, K., and R. Strand. 2011. Conversations about responsible nanoresearch. *NanoEthics* 5(1): 99–113.

Malsch, I., and Hvidtfelt Nielsen, K. 2009. *Individual and collective responsibility for nanotechnology.* First annual report on ethical and social aspects of the ObservatoryNano project.

Mancuso, S. 2008. Ma è oscurantismo? No, tutela dell'uomo. *Corriere della Sera*, December 9.

McCarthy, E., and C.M. Kelty. 2010. Responsibility and nanotechnology. *Social Studies of Science* 40(3): 405–432.

Miano, F. 2009. *Responsabilità*. Napoli: Guida Editori.

Milani, P. 2011. Esperti internazionali al Museo della scienza; Le nanotecnologie sono un 'miracolo' o una 'minaccia'? *Corriere della Sera*, October 11.

Paterlini, M. 2011. I miei nanotubi vi guarirannò' Dopo le missioni nell'organismo, si degradano grazie a un enzima scoperto nei globuli bianchi. *La Stampa*, May 11.

Pellizzoni, L. 2010. Risk and responsibility in a manufactured world. *Science and Engineering Ethics* 16(3): 463–479.

Ricci, P. 2010. Al via la cittadella della scienza. *La Repubblica*, March 16.

Rizzato, S. 2011. Nanotech: si affina un'arma decisiva. *La Stampa*, November 9.

Roco, M., and W.S. Bainbridge. 2001. Introduction. In *Societal implications of nanoscience and nanotechnology*, ed. M. Roco and W.S. Bainbridge, 1–19. Arlington: National Science Foundation.

Scheufele, D.A., Corley, E.A., Shih, T., Dalrymple, K.E. and S.S. Ho. 2009. Religious beliefs and public attitudes toward nanotechnology in Europe and the United States. *Nature Nanotechnology* 4: 91–94.

Strydom, P. 1999. The challenge of responsibility for sociology. *Current Sociology* 47(3): 65–88.

Suchman, M.C. 2001. Envisioning life on the nano-frontier. In *Societal implications of nanoscience and nanotechnology*, ed. M. Roco and W.S. Bainbridge, 211–216. Arlington: National Science Foundation.

Tolles, W.M. 2001. National security aspects of nanotechnology. In *Societal implications of nanoscience and nanotechnology*, ed. M. Roco and W.S. Bainbridge, 161–187. Arlington: National Science Foundation.

Epilogue: Nanotechnology Beyond Nanotechnologies

Chapter 12
Responsible Research and Innovation: An Emerging Issue in Research Policy Rooted in the Debate on Nanotechnology

Armin Grunwald

12.1 Introduction and Overview

The advance of science and technology has been accompanied by debates in society on issues of risks and chances, potentials and side effects, control and responsibility for decades. Approaches such as technology assessment (Grunwald 2009a), social shaping of technology (Yoshinaka et al. 2003), responsibility ethics (Durbin and Lenk 1987) and value sensitive design (van de Poel 2009) have been developed and are practiced to a certain extent. All of them have a specific focus, particular theoretical foundations, different rationales, and were conceptualized for meeting differing constellations. All of them also show strengths and weaknesses and specific limitations to application. Therefore, research towards new and more comprehensive concepts is still ongoing.

Responsible Research and Innovation (RRI) is a rather new element of research policy and technology governance (Grunwald 2011a; von Schomberg 2012). Its emergence (Siune et al. 2009) reflects the diagnosis that available approaches to shape science and technology still do not meet all of the far-ranging expectations. The hope behind the international RRI debate is that new – or further-developed – approaches could add considerably to existing approaches such as technology assessment and engineering ethics. And expectations are high as can be seen from the fast career of the RRI notion. Today RRI is one of the major elements of the new European research framework programme Horizon2020.

In this chapter I will first introduce some basics of Responsible Research and Innovation, focusing on the integrative character of this approach (Sect. 12.2). Because its keyword 'responsible' needs clarification I then will propose a three-dimensional understanding of responsibility which fits very well to the

A. Grunwald (✉)
Institute for Technology Assessment and Systems Analysis (ITAS),
Karlsruhe Institute of Technology (KIT), Karlsruhe, Germany
e-mail: armin.grunwald@kit.edu

S. Arnaldi et al. (eds.), *Responsibility in Nanotechnology Development*, The International Library of Ethics, Law and Technology 13, DOI 10.1007/978-94-017-9103-8_12,
© Springer Science+Business Media Dordrecht 2014

integrative character of RRI (Sect. 12.3). Of particular interest is the relation of RRI and the debate on nanotechnology and society. Historically, the RRI notion emerged in the context of this debate going back to the National Nanotechnology Initiative of the U.S. It was quickly taken up by European nanotechnology policy, e.g. in the debate on the Code of Conduct for nanotechnology research and development set in practice by the European Parliament. My intention is to demonstrate the parallel development of nano-ethics on the one side, and the debate on Responsible Research and Innovation, on the other, as well as at least partially to attempt to understand this parallel development (Sect. 12.4). The thesis is that many of the experiences made in the debate on ethics of nanotechnology contributed to proposing and shaping the RRI concept, in particular experiences with nanotech as a new and emerging technology promising revolutionary potential but also unclear risk. The debate on nanotech & ethics thus developed to a 'model' of dealing responsibly with new and emerging sciences and technologies.[1]

12.2   Responsible Research and Innovation

In the RRI context different notions are used such as Responsible Innovation, Responsible Research and Development, and Responsible Research and Innovation. In this chapter, the notion of 'Responsible Research and Innovation' (RRI) will be used because it allows integrating the perspective on scientific research as source of technological innovation on the one hand, and the view on innovation referring to processes in 'real-world' society including the economy, on the other.

The ideas of 'responsible research' in the scientific-technological advance and of 'responsible innovation' in the field of new products, services and systems have been discussed for some years now with increasing intensity. The RRI concept has emerged mainly in connection with a large variety of new technologies regarded as key technologies, such as synthetic biology, nanotechnology, new internet technologies, robotics, geoengineering, etc. However, the motivation to speak of responsible research and innovation goes back to large-scale national programs to conduct R&D on nanotechnology. The US National Nanotechnology Initiative (NNI) adopted a strategic goal of 'responsible development':

Responsible development of nanotechnology can be characterized as the balancing of efforts to maximize the technology's positive contributions and minimize its negative consequences. Thus, responsible development involves an examination both of applications and of potential implications. It implies a commitment to develop and use technology to help meet the most pressing human and societal needs, while making every reasonable effort to anticipate and mitigate adverse implications or unintended consequences. (National Research Council 2006, p. 73)

---

[1] This chapter consists of an integration and further development of other work of the author (Grunwald 2011a, b, 2012a).

Other actors in the field of research policy quickly followed. The UK Engineering and Physical Sciences Research Council published a study on responsible innovation for nanotechnology in the field of carbon capture. The Netherlands organized a 'national dialogue' on nanotechnology requesting that further development in nanotechnology should be 'responsible' (Guston et al. 2014). The European Union adopted a code of conduct for nanoscience and nanotechnology (N&N) research (EC 2008) referring to research and development but also to public understanding and the importance of precaution. It also links responsibility reflection to governance (following Siune et al. 2009, p. 32): the guidelines "are meant to give guidance on how to achieve good governance".

Good governance of N&N research should take into account the need and desire of all stakeholders to be aware of the specific challenges and opportunities raised by N&N. A general culture of responsibility should be created in view of challenges and opportunities that may be raised in the future and that we cannot at present foresee (EC 2008; following Siune et al. 2009, p. 32).

Nanotechnology has attracted all this attention, because it is an example of a technology that is known for its potential high stakes, uncertainty and possible adverse effects.[2] The purpose of these endeavors is to enhance the possibilities that technology will help to improve the quality of human life, that possible non-intended side-effects will be discovered as early as possible in order to enable society to prevent or compensate them, and that, accordingly, those technologies and innovations will be socially accepted.

This rationale is well-known from the field of technology assessment (TA; Grunwald 2009a), in particular from Constructive TA (Rip et al. 1995). The Control Dilemma (Collingridge 1980), however, emphasizes that shaping technology to optimally harvest intended and to avoid non-intended effects is an ambitious task being in danger of either coming too late or too early. Facing this Dilemma, the conceptual development of major parts of TA over approximately the last 10 years may be characterized as an 'upstream movement' to the early stages of technology development (e.g. van der Burg and Swierstra 2013). The expectation was and still is that giving shape to technology should be possible also in the case of only little knowledge being available about applications and usage of the technology under consideration. The Control Dilemma shall be circumvented by various approaches (Liebert and Schmidt 2010). The fields of technology considered belong to new and emerging sciences and technologies (NEST) such as nanotechnology, nano-biotechnology and synthetic biology. They thus show a strong "enabling character" leading to a variety of possible applications in different areas which are extremely difficult to anticipate. This situation makes it necessary to shape any reflective activity such as TA as an *accompanying process* referring to the ethical, social, legal and economic issues at stake. This process character which is well-known from the field of TA (van Eindhoven 1997) is very attractive to RRI.

---

[2] Because of this history we will give more emphasis to the emergence of the RRI idea in the field of nanotech later on in this chapter (Sect. 12.4).

Consequently, the RRI definition recently proposed by René von Schomberg breathes the spirit of technology assessment in this sense (see also von Schomberg 2007) because it basically introduces RRI as a *process*, enriched by normative elements derived from the responsibility issue:

Responsible Research and Innovation is a transparent, interactive process by which societal actors and innovators become mutually responsive to each other with a view to the (ethical) acceptability, sustainability and societal desirability of the innovation process and its marketable products (in order to allow a proper embedding of scientific and technological advances in our society) (von Schomberg 2012)

Responsible Research and Innovation thus adds explicit ethical reflection to the procedural "upstream movement" of TA and involves the ethics of engineering and technology (Grunwald 2013) as the second major root of RRI. RRI brings together TA with its experiences on assessment procedures, actor involvement, participation, foresight and evaluation with engineering and technology ethics, in particular under the framework of responsibility (Jonas 1984; Durbin and Lenk 1987). This integration overcomes the separation of ethics and TA which led to heavy discussions in the 1990s (Grunwald 1999).

A further integration concerns the relation of ethics and TA on the one side, and actor constellations and contexts of deliberation and decision-making, on the other. Because RRI applies a 'make-perspective' the socio-political dimension of the processes under consideration must be taken into account – and this leads to the necessity of involving social sciences, in particular from the field of STS studies (science, technology and society). RRI unavoidably requires a more intense inter- and trans-disciplinary cooperation between engineering, social sciences, and applied ethics. Thus, the novelty of RRI mainly consists of this integrative approach (Grunwald 2011a) which also contributes to its attractiveness and is source of most of the far-ranging expectations related to RRI.

An operable example of what RRI could mean in practice is the research program "Responsible Innovation – Ethical and Societal Exploration of Science and Technology" (MVI, following its Dutch name) of the Dutch Organization for Scientific Research (NWO). The MVI program – which is among the earliest manifestations of RRI – "focuses on technological developments which we can expect will have an impact on society. On the one hand, those developments concern new technologies (such as ICT, nanotechnology, biotechnology and cognitive neuroscience), and on the other, technological systems in transition (for example agriculture and healthcare). The MVI contributes to responsible innovation by increasing the scope and depth of research into societal and ethical aspects of science and technology" (NWO 2013). The projects funded under this program have to demonstrate a 'make'-perspective beyond mere scientific research: "Projects for research into ethical and societal aspects of concrete technological developments must always have a 'makeable perspective'. In other words, they must not only lead to an analysis and an improved understanding of problems, but also result in a 'design perspective' – in the broadest sense, including institutional arrangements" (NWO 2013).

The MVI program started in 2009 funding 15 projects in the first round (see NWO 2013 for the list of projects and short descriptions). One example is the project *New economic dynamics in small producers' clusters in northern Vietnam – Institutions and responsible innovation with regard to poverty alleviation* on the analysis and enhancement of the value-added chains of local producers. It "builds further on the research outcome by exploring the potential importance of these specific technological cases for poverty reduction in developing countries, thus whether the innovations could be labelled as 'responsible innovations'". Vietnam offers a particularly interesting research context since the innovations of poor small producers are based on private initiatives with an institutional environment in transition and aims

- 'to understand the concept of "responsible innovation" and its valorization in small producers' clusters in northern Vietnam';
- 'to explain the multi-level institutional framework enabling and facilitating the small producers to innovate';
- 'to assess how the institutional framework interacts with small producers' economic behaviour through incentives'.

This description clearly shows that research in the framework of RRI is not an end in itself but rather a means for analyzing and then improving the conditions of local life in the region considered. A Valorisation Panel – obligatory to all of the MVI projects – takes care that the 'make'-perspective is observed in conducting the projects.

This example shows some of the aspects that the Responsible Research and Innovation approach emphasizes when compared to existing approaches such as TA and engineering ethics (Grunwald 2011a):

- 'shaping innovation' complements or even replaces the former slogan 'shaping technology' which characterised the approach by social constructivist ideas to technology (Bijker and Law 1994). This shift reflects the insight that it is not technology as such which influences society and therefore should be shaped according to human needs, expectation and values, but it is innovation by which technology and society interact;
- there is a closer look on societal contexts of new technology and science. RRI can be regarded as a further step towards taking the demand-pull perspective and social values in shaping technology and innovation more serious;
- instead of expecting distant observation following classical paradigms of science there is a clear indication for intervention into the development and innovation process: RRI projects shall 'make a difference' not only in terms of research but also as interventions into the 'real world';
- based on earlier experiences with new technologies such as genetic engineering and with corresponding moral and social conflicts, a strong incentive is to 'get things right from the very beginning' (Roco and Bainbridge 2001);
- user involvement, stakeholder involvement, and citizen involvement into the research and innovation processes are regarded as an important approach to better integrate societal needs and perspectives on the one hand, and technology and innovation, on the other (von Schomberg 2012).

Thus, Responsible Research and Innovation can be regarded as a further development, even a radicalisation of the well-known post-normal science (Functowitz and Ravetz 1993) being even closer to social practice, being prepared for intervention and for taking responsibility for this intervention.

Currently, an international community on RRI issues is organizing itself. The MVI program mentioned above is one of its origins, others coming from the related fields of Value Sensitive Design, Engineering Ethics, Technology Assessment, and Anticipatory Governance. The first institutional manifestations took place in the Seventh Framework Programme of the European Union where some projects aiming at making RRI more operable were resp. still are being funded. In the new Horizon2020 program RRI is a major and cross-cutting issue. The International Journal of Responsible Innovation which will be launched in 2014 shall provide a platform for further developing RRI in theory, methodology, and practice:

> It is our hope that the *Journal of Responsible Innovation* will help communicate and deliberate such work, which not only involves scholars and teachers from a variety of disciplines and fields including natural science and engineering, sociology, anthropology, moral philosophy, political science, science and technology studies, but also practitioners in such professional areas as technology assessment, management and strategy, research funding, and science and innovation policy. The *Journal of Responsible Innovation* will help manifest and broaden that network by providing a platform to articulate and discuss the many unsolved questions surrounding responsible innovation, and by inviting new and surprising perspectives of scholars and practitioners who take an interest in reflecting on and debating this theme (Guston et al. 2014).

At the scientific side the foundation of the trans-atlantic 'Virtual Institute of Responsible Innovation' (VIRI) should be mentioned as a major step to giving shape to the institutional landscape of RRI. The VIRI was initiated and is coordinated by Arizona State University and funded by the U.S. National Science Foundation. Thus, the RRI concept gave and gives rise to form an international community of scholars, researchers, and practitioners to further develop and make the concept operable.

12.3   Three Dimensions of Responsibility

In spite of the many debates on responsibility in the fields of science and technology (e.g. Durbin and Lenk 1987) the very meaning of this notion is still vague. There are concerns that responsibility is often used as a merely rhetorical phrase with a character of appealing or even preaching to people but with more or less only little practical consequences. These concerns should be taken seriously, and effort should be spent to work on strategies to overcome them. It will be shown in this Section[3] that, in order to make the concept of responsibility work in the framework of RRI, three dimensions have to be considered: the empirical dimension, the epistemic

---

[3] This Section follows closely the argumentation and formulation given in Grunwald (2012a, b).

dimension, and the ethical dimension. The Precautionary Principle (Harremoes et al. 2002) can be regarded, against this background, as an operative example for making responsibility work by involving all three dimensions (von Schomberg 2005).

What 'responsible' in a specific context means and what distinguishes 'responsible' from 'irresponsible' or less responsible innovation is difficult to determine. The distinction will strongly depend on values, rules, and customs etc. and vary according to different context conditions. The notion of RRI as such does not give orientation how to meet this challenge. Rather, a more in-depth look at the concept of responsibility is required (Grunwald 2012a, b, cf. further references given there). Responsibility in social practice is regarded the result of an *assignment process*. Assignments and attributions of responsibility affect concrete actors in concrete constellations – this observation motivates putting emphasis on the empirical constellation of responsibility. The aim of assigning responsibility in RRI cases is, generally speaking, contributing to establishing a 'good practice' in the respective field. A four-place reconstruction generally seems to be suitable for discussing issues of responsibility in scientific and technical progress (Grunwald 2012a, b):

- *someone* (an actor, e.g. a nanotech researcher) assumes responsibility or is made responsible (responsibility is assigned to her/him) for
- *something* such as the results of actions or decisions, e.g. for avoiding adverse health effects of nano-materials, relative to
- *rules and criteria* which orientate responsible from less responsible or irresponsible action, and relative to the
- *knowledge available* about the impacts and consequences of the action or decision under consideration, including also meta-knowledge about the epistemological status of that knowledge and uncertainties involved.

Though the first two places are, in a sense, trivial in order to make sense of the word responsible, they indicate the fundamental social context of assigning responsibility as an empirically observable process among social actors. The third and fourth places open up further essential dimensions of responsibility. The dimension of rules and criteria reflects on principles, norms and values being decisive for the judgment of whether a specific action or decision is regarded responsible or not. Analysis and reflection of these normative elements constitute the *ethical dimension* of responsibility. The knowledge available and its quality, including all the uncertainties, form its *epistemic dimension*. Relevant questions and challenges in the RRI context arise in all of these three dimensions and thus must be considered together (following Grunwald 2012a, b):

- the *empirical dimension* of responsibility seriously considers that the attribution of responsibility is an act of specific actors which affects others. It refers to the basic social constellation of assignment processes. Assignment of responsibility must, on the one hand, take into account the possibilities of actors to influence their actions and decisions in their respective fields. Issues of accountability and power must be taken into account. On the other, attributing responsibilities has an impact on the governance of that field. Shaping that *governance* is the ultimate goal of debating issues of assigning and distributing responsibility *ex ante*. Relevant questions are: How are capabilities, influence, and power to act, as well as decisions taken in the field, considered? Which social groups are affected, and should they help determine the distribution of responsibility? Do the questions under consideration concern issues to be debated at the *polis* or can they be delegated to groups or subsystems? What consequences would a particular distribution of responsibility have for the governance of the respective field, and would it be in favor of desired developments?

- the *ethical dimension* of responsibility is reached when the question is posed for criteria and rules for judging actions and decisions under consideration as responsible or irresponsible, or for helping to find out how actions and decisions could be designed to be (more) responsible. Insofar as normative uncertainties arise (Grunwald 2012b), e.g. because of ambiguity or moral conflicts, ethical reflection on these rules and their justifiability is needed. Relevant questions are: What criteria allow distinguishing between responsible and irresponsible actions and decisions? Is there consensus or controversy on these criteria among the relevant actors? Can the actions and decisions in question (e.g., about the scientific agenda or about containment measures to prevent bio-safety problems) be regarded as responsible with respect to the rules and criteria?

- the *epistemological* dimension asks for the knowledge about the subject of responsibility and its epistemological status and quality. This is a particularly relevant issue in debates on scientific responsibility because, frequently, statements about the impact and consequences of science and new technology show a high degree of uncertainty. The comment that nothing else comes from "mere possibility arguments" (Hansson 2006) is an indication that, in debates over responsibility, it is essential that the status of the available knowledge about the accountable future is determined and is critically reflected upon from an epistemological point of view. Relevant questions are: What is really known about prospective subjects of responsibility? What could be learned through more research, and which uncertainties are pertinent? How can different uncertainties be qualified and compared to each other? And what is at stake if worse comes to worst?

Debates over responsibility in technology and science frequently focus exclusively on the *ethics* of responsibility (Durbin and Lenk 1987). However, regarding the analysis given so far, this is only part of the field and neglects the empirical as well as the epistemological dimension of responsibility. It seems that the familiar criticisms towards responsibility reflections (see above) of being simply appellative, of epistemological blindness, and of being politically naïve, are related to narrowing responsibility to its ethical dimension (Grunwald 2012a). The brief theoretical analysis given above showed that the issue of responsibility is not only one of abstract ethical judgment but necessarily includes issues of concrete social contexts. Governance factors must be treated empirically as well as the issue of the epistemological quality of the knowledge available. Meeting those criticisms and making the notion of responsibility work is claimed to be possible by considering the EEE dimensions of responsibility together.

This result matches exactly what has been said in Sect. 12.2 on the integrative character of the RRI approach. It is crucial to understand that RRI is much more than ethics of responsibility. It rather includes procedures of involvement and of deliberation for determining an adequate realization of distribution of responsibility in the respective empirical constellation and procedures and criteria of taking the epistemic quality of the knowledge available into consideration (e.g. Pereira et al. 2007). While high emphasis is currently spent on the procedures of involvement (of stakeholders, citizens, users etc.) in RRI the epistemic dimension seems to be underdeveloped so far (Grunwald 2014).

12.4   Nano-ethics – The Roots of the RRI Approach?

Nanotechnologies were perceived as seemingly non-risk technologies for a long time. Public perception in the 1990s was low. The prefix 'nano', however, and this is a strong indication of a positive perception, was used in the media – not in mass media but, for instance, in science magazines – as a synonym for 'good' science and 'smart' technology. This situation changed radically in 2000. The positive utopias of nanotechnologies, based on a technical access to 'the small', were inverted to horror scenarios, based on the same 'small' technologies (Joy 2000). The ambivalence of technology-based visions became obvious. The public risk debate on nanotechnology emerged around issues of visionary and more speculative developments. Topics like 'grey goo', 'nanobots' and 'cyborgs' became well-known to many people within a few months (Schmid et al. 2006, chap. 5). Concerned groups began to think about analogies and parallels between nanotechnologies and technology lines with a specific history in the public risk debate: nuclear technology and biotechnology (ETC Group 2003). Newspapers and reassurance companies put nanotechnology in the category of risky technologies.[4]

The emergence of the risk issue in combination with the fact of having practically no knowledge available about side effects of nanotechnology led to severe irritation and to a kind of helplessness at the early stage of that debate. My thesis explained below is that the quick career of the notion of responsibility in research and innovation policy is related with this situation. Statements from that time waver between an optimistic 'wait-and-see' strategy (Gannon 2003) on the one hand, and strict precautionary and sometimes 'alarmist' approaches on the other:

The new element with this kind of loss scenario is that, up to now, losses involving dangerous products were on a relatively manageable scale, whereas, taken to extremes, nanotechnology products can even cause ecological damage which is permanent and difficult to contain. What is therefore required for the transportation of nanotechnology products and processes is an organisational and technical loss

---

[4] This Section goes in parts back to the review of the history of the nanotech and ethics debate given in Grunwald (2011b).

prevention programme on a scale appropriate to the hazardous nature of the products (Munich Re 2002, p. 13).

The still most famous position on nanoparticle regulation is probably the postulate of the ETC Group for a moratorium:

> At this stage, we know practically nothing about the possible cumulative impact of human-made nanoscale particles on human health and the environment. Given the concerns raised over nanoparticle contamination in living organisms, the ETC Group proposes that governments declare an immediate moratorium on commercial production of new nanomaterials and launch a transparent global process for evaluating the socio-economic, health and environmental implications of the technology (ETC Group 2003, p. 72).

The ETC work gave a significant push to nanotechnology regulatory debates in many countries, but also increased the fears on the side of nanotech researchers of a broad public front of rejection and protest. The Center for Responsible Nanotechnology (CRN) was specifically founded to contribute to meeting those challenges brought up by nanotechnology:

The mission of CRN is to: (1) raise awareness of the *benefits*, the *dangers*, and the possibilities for responsible use of advanced nanotechnology; (2) expedite a *thorough examination* of the environmental, humanitarian, economic, military, political, social, medical, and ethical implications of molecular manufacturing; and (3) assist in the *creation and implementation* of wise, comprehensive, and balanced plans for responsible worldwide use of this transformative technology (CRN 2014).

Compared to the ETC position mentioned above CRN published a completely different but also far-reaching recommendation also in 2003. It aimed at 'containing' nanotech research breathing the same spirit of uneasiness, high uncertainty, and irritation. CRN has identified several sources of risk from MNT (molecular nanotechnology), including arms races, gray goo, societal upheaval, independent development, and programmes of nanotech prohibition that would require a violation of human rights. It appears that the safest option is the creation of one – and only one – molecular nanotechnology programme and the widespread but restricted use of the resulting manufacturing capability (Phoenix and Treder 2003, p. 4).

This containment strategy would imply a secret and strictly controlled nanotech development, which seems to be unrealistic and unsafe as well as undemocratic. Furthermore, this recommendation is irritating regarding the ideal of an open scientific community.

These different proposals have enriched (and heated) public and scientific debate. Seen from today's perspective (Grunwald 2011b), they document a very specific situation. Nanotechnology, still in a very early state of development, found itself the subject of lively public debate. While high expectations of benefits still dominated, there was no reliable knowledge about the possible side effects of nanotechnology. Against this background, it is understandable that the first years of the nanotech ethics and risk debate were largely based on mere suspicion, irritations and uncertainties rather than on knowledge-based and rational deliberation.

In exactly this situation first indications of the later RRI approach emerged (see Sect. 12.2). The career of the notion of responsibility – to be honest, it was the second career, the first followed the work by Hans Jonas (1984) – in the context of this early debate on nanotech & society did not happen by chance, so the thesis of this Section. Usually, the emergence of new technologies is accompanied by technology assessment conducting, for example, life cycle assessments (LCA) in order to understand the impacts of the technology under consideration over the entire lifetime. Or ethical reflection is made, putting the new technology into its envisaged societal context and asking for morally relevant impacts and consequences. However, in the case of the early debate on nanotech there was almost no knowledge available for doing an LCA or taking development lines based on nanotech as subject to an ethical deliberation. Neither TA in its familiar meaning nor ethical reflection in the sense of applied ethics was possible because of lack of knowledge.[5]

One approach to cope with this situation was to conduct ELSA studies (ethical, legal, and social aspects) carried out by expert groups and encompassing issues such as equity, governance, participation and access (e.g. Nanoforum 2004). The improvisatory way these studies were performed was a consequence of the situation described above. It was an approach to provide some kind of orientation in a situation where provision of orientation by established approaches such as TA and ethics seemed to be impossible. The result was modest but better than nothing: lists of ethical, legal, social and perhaps other aspects including the recommendation that the elements of those lists should be observed and investigated more in depth in the further development of nanotechnology. Indeed, these lists later on were developed further to a canonical set of ethical aspects of nanotechnology (Grunwald 2011c, 2012b). In this sense, the ELSI period of addressing nanotechnology and social issues (approx. 2004–2006) could be regarded as an exploratory stage that contributed decisively to the agenda-setting and structuring of the emerging field of nano-ethics.

Furthermore, the word "responsible" was added to nanotechnology and its research and development (e.g. by National Research Council 2006) – probably without having a clear impression what this should mean in detail. The word "responsible" in the early nanotech debate seems to have been a void notion covering the situation of high uncertainty in combination with the conviction that nanotech would have some revolutionary impacts on the future. At that time the usage of the word "responsible" simply might have helped to accept a statement like "Tremendous transformative potential comes with tremendous anxieties" (Nordmann 2004, S. 4). The solution was, psychologically simple, adding the word "responsible" to nanotech in order to make peace with "tremendous anxieties".

However, this approach was more than psychology. The frequent usage of the notion of "responsible" called for making it operable which was done in several respects in the next phase of the nanotech debate:

---

[5] The criticism on the 'speculative nano-ethics' (Nordmann 2007) was, in a sense, a rather late reflection on this situation.

- 'responsible' is a notion which is traditionally related with the behavior of individual persons. Thus it was consequential to call for a "Code of Conduct" for all people involved in nanotech research and development. Actions under such a Code (realized in the European Union, EC 2008) should *per se* be responsible.
- looking more from a TA perspective it was obvious that though TA through LCA was not possible there could be other ways of providing orientation. The approach of Vision Assessment (Grunwald 2009b; Ferrari et al. 2012) was a result of this line of reflection
- furthermore, following former experiences in the field of the Precautionary Principle (von Schomberg 2005; Weckert and Moor 2007), it became clear that analytical approaches must be complemented by social procedures involving actors, stakeholders, citizens etc.

All these elements are currently being regarded part of the RRI approach (Grunwald 2011a; von Schomberg 2012). This observation legitimizes regarding the early nanotech debate and its consequences as the roots of the RRI movement we have today. Even a bit stronger, the hypothesis may be proposed that the early nanotech debate and the consequences it had in terms of developing more reflective and multi-faceted approaches to deal with novel situations may be taken as a model for coming debates in the NEST field.

12.5 Perspectives and Challenges

The terms of responsible development, responsible research and responsible innovation are highly integrative because they cover issues of engineering ethics, participation, technology assessment, anticipatory governance and science ethics. In this sense, responsible development and innovation is a rather new umbrella term (von Schomberg 2012) promoting new accentuations in the co-evolution of technology and society characterized by

- involving ethical and social issues more directly in the innovation process by integrative approaches to development and innovation;
- bridging the gap between innovation practice, engineering ethics, technology assessment, governance research and social sciences (STS);
- giving new shape to innovation processes and to technology governance according to responsibility reflections in all of its three dimensions – empirical, ethical, and epistemological;
- in particular, making the distribution of responsibility among the involved actors as transparent as possible;
- integrating a heterogeneous set of tools, methods, procedures, and approaches to be "fit for purpose";
- covering the entire innovation process from the very beginning (visions, expectations) up to specific and contextual pathways of innovation under political circumstances and in an economic situation of competition;

- supporting "constructive paths" of the co-evolution of technology and the regulative frameworks of society.

However, it is important to point out that the model of integrated research including its own ethical and responsibility reflection also harbours problems (following Grunwald 2011b). The independence of reflection can be threatened especially if the necessary distance to the technical developments and those working on them is lost. Inasmuch as assessment issues become part of the development process and would identify itself with the technical success, there might be an accusation that its acceptance was "purchased" or that it was nothing but grease in the process of innovation. Strategies of dealing with such possible developments should be developed and could include means such as careful monitoring activities and a strong role of external review processes. It will be a task for the respective emerging research community around the RRI issue to take care but also the responsible funding agencies should be aware of this challenge.

This challenge is not absolutely new. It rather accompanies problem-oriented research such as technology assessment and applied ethics over time. Characteristic is the twofold mission: these approaches have, on the one hand, to stick to the claim for scientific excellence and independence, and, on the other, to 'make a difference' in the real world and to create impacts on ongoing processes of development and innovation.

References

Bijker, W., and J. Law (eds.). 1994. *Shaping technology/building society*. Cambridge, MA: The MIT Press.

Collingridge, D. 1980. *The social control of technology*. London: Pinter.

CRN – Center for Responsible Nanotechnology. 2014. https://www.google.de/#q=crn+center+for+responsible+nanotechnology. Accessed 22 Jan 2014.

Durbin, P., and H. Lenk (eds.). 1987. *Technology and responsibility*. Dordrecht: Reidel Publishing.

EC – European Commission. 2008. *Commission recommendation on a code of conduct for responsible nanosciences and nanotechnologies research*. 2008/424 final, February 7, 2008. Brussels: European Commission.

ETC Group. 2003. *The big down. From genomes to atoms. Atomtech: Technologies converging at the nano-scale*. Ottawa: ETC Group.

Ferrari, A., C. Coenen, and A. Grunwald. 2012. Visions and ethics in current discourse on human enhancement. *NanoEthics* 6(3): 215–229.

Functowitz, S., and J. Ravetz. 1993. The emergence of post-normal science. In *Science, politics and morality*, ed. R. von Schomberg, 85–124. Amsterdam: North Holland.

Gannon, F. 2003. Nano-nonsense. *EMBO Reports* 4: 1007.

Grunwald, A. 1999. Technology assessment or ethics of technology? Reflections on technology development between social sciences and philosophy. *Ethical Perspectives* 6: 170–182.

Grunwald, A. 2009a. Technology assessment: Concepts and methods. In *Philosophy of technology and engineering sciences*, ed. A. Meijers, 1103–1146. Amsterdam: Elsevier.

Grunwald, A. 2009b. Vision assessment supporting the governance of knowledge – The case of futuristic nanotechnology. In *The social integration of science. Institutional and epistemological aspects of the transformation of knowledge in modern society*, ed. G. Bechmann, V. Gorokhov, and N. Stehr, 147–170. Berlin: Edition Sigma.

Grunwald, A. 2011a. Responsible innovation: Bringing together technology assessment, applied ethics, and STS research. *Enterprise and Work Innovation Studies* 7: 9–31.

Grunwald, A. 2011b. Ten years of research on nanotechnology and society – Outcomes and achievements. In *Quantum engagements: Social reflections of nanoscience and emerging technologies*, ed. T.B. Zülsdorf, C. Coenen, A. Ferrari, U. Fiedeler, C. Milburn, and M. Wienroth, 41–58. Heidelberg: AKA GmbH.

Grunwald, A. 2011c. Chances and risks of nanotechnology. In *Handbook of nanophysics. Nanomedicine and nanorobotics*, ed. K.D. Sattler, 13–16. Boca Raton: CRC Press.

Grunwald, A. 2012a. Synthetic biology: Moral, epistemic and political dimensions of responsibility. In *Proceed with caution? – Concept and application of the precautionary principle in nanobiotechnology*, ed. R. Paslack, J.S. Ach, B. Luettenberg, and K. Weltring, 243–259. Münster: LIT Verlag.

Grunwald, A. 2012b. *Responsible nanobiotechnology. Philosophy and ethics*. Singapore: Pan Stanford.

Grunwald, A. (ed.). 2013. *Handbuch Technikethik*. Stuttgart: Metzler.

Grunwald, A. 2014. *On the epistemological dimension of responsible research and innovation* (Submitted).

Guston, D., E. Fisher, A. Grunwald, R. Owen, T. Swierstra, and S. van der Burg. 2014. Responsible innovation. Motivations for a new journal. *Journal of Responsible Innovation* 1(1): 1–8.

Hansson, S.O. 2006. Great uncertainty about small things. In *Nanotechnology challenges – Implications for philosophy, ethics and society*, ed. J. Schummer and D. Baird, 315–325. Singapore: World Scientific Publishing.

Harremoes, P., D. Gee, M. MacGarvin, A. Stirling, J. Keys, B. Wynne, and S. Guedes Vaz (eds.). 2002. *The precautionary principle in the 20th century. Late lessons from early warnings*. London: Earthscan.

Jonas, H. 1984. *The imperative of responsibility*. Chicago: University of Chicago Press.

Joy, B. 2000. Why the future does not need us. *Wired Magazine*, pp. 238–263.

Liebert, W., and J. Schmidt. 2010. Collingridge's dilemma and technoscience. *Poiesis & Praxis* 7: 55–71.

Munich Re. 2002. *Nanotechnology – What is in store for us?* www.anet.co.il/anetfiles/files/241M.pdf. Accessed 24 Jun 2014.

Nanoforum. 2004. *Nanotechnology. Benefits, risks, ethical, legal, and social aspects of nanotechnology*. http://www.nanowerk.com/nanotechnology/reports/reportpdf/report3.pdf. Accessed 31 Jan 2014.

National Research Council. 2006. *A matter of size: Triennial review of the national nanotechnology initiative*. Washington, DC: National Academies Press.

Nordmann, A. 2004. *Converging technologies – Shaping the future of European societies*. Luxembourg: Office of the Official Publications of the European Communitie.

Nordmann, A. 2007. If and then: A critique of speculative nanoethics. *NanoEthics* 1: 31–46.

NWO – Dutch Organization of Research. 2013. *Homepage of the MVI program*. http://www.nwo.nl/en/research-and-results/programmes/responsible+innovation. Accessed 21 Dec 2013.

Pereira, A.G., R. von Schomberg, and S. Funtowicz. 2007. Foresight knowledge assessment. *International Journal of Foresight and Innovation Policy* 4: 65–79.

Phoenix, C., and Treder, M. 2003. *Applying the precautionary principle to nanotechnology*. http://www.crnano.org/Precautionary.pdf. Accessed 31 Jan 2014.

Rip, A., T. Misa, and J. Schot (eds.). 1995. *Managing technology in society*. London: Pinter Publishers.

Roco, M.C., and W.S. Bainbridge (eds.). 2001. *Societal implications of nanoscience and nanotechnology*. Boston: Kluwer Academic Publishers.

Schmid, G., H. Ernst, W. Grünwald, A. Grunwald, H. Hofmann, P. Janich, H. Krug, M. Mayor, W. Rathgeber, B. Simon, V. Vogel, and D. Wyrwa. 2006. *Nanotechnology – Perspectives and assessment*. Berlin: Springer.

Siune, K., E. Markus, M. Callon, U. Felt, A. Gorski, A. Grunwald, A. Rip, V. de Semir, and S. Wyatt. 2009. *Challenging futures of science in society*. Report of the MASIS expert group. Brussels: European Commission.

van de Poel, I. 2009. Values in engineering design. In *Philosophy of technology and engineering sciences*, ed. A. Meijers, 973–1006. Amsterdam: Elsevier.

van der Burg, S., and T. Swierstra (eds.). 2013. *Ethics on the laboratory floor*. Basingstoke: Palgrave Macmillan.

van Eindhoven, J. 1997. Technology assessment: Product or process? *Technological Forecasting and Social Change* 54: 269–286.

von Schomberg, R. 2005. The precautionary principle and its normative challenges. In *The precautionary principle and public policy decision making*, ed. E. Fisher, J. Jones, and R. von Schomberg, 141–165. Cheltenham: Edward Elgar.

von Schomberg, R. 2007. *From the ethics of technology towards an ethics of knowledge policy and knowledge assessment*. Working document from the European Commission. https://europa.eu/sinapse/sinapse/index.cfm?&fuseaction=lib.attachment&lib_id=5F13003C-C1F1-A556-AD4AF9FD76D4BD84&attach=LIB_DOC_EN. Accessed 31 Jan 2014.

von Schomberg, R. 2012. Prospects for technology assessment in the 21st century: The quest for the "right" impacts of science and technology. An outlook towards a framework for responsible research and innovation. In *Technikfolgen abschätzen lehren*, ed. M. Dusseldorp and R. Beecroft, 371–388. Opladen: Springer.

Weckert, J., and J. Moor. 2007. The precautionary principle in nanotechnology. In *Nanoethics – The ethical and social implications of nanotechnology*, ed. F. Allhoff, P. Lin, J. Moor, and J. Weckert, 133–146. Hoboken: Wiley.

Yoshinaka, Y., C. Clausen, and A. Hansen. 2003. The social shaping of technology: A new space for politics? In *Technikgestaltung: zwischen Wunsch oder Wirklichkeit*, ed. A. Grunwald, 117–131. Berlin: Springer.

Index

S. Arnaldi et al. (eds.), *Responsibility in Nanotechnology Development*, The International
Library of Ethics, Law and Technology 13, DOI 10.1007/978-94-017-9103-8,
© Springer Science+Business Media Dordrecht 2014